T0388754

Sustainable Land Management for NEOM Region

Mashael M. Al Saud

# Sustainable Land Management for NEOM Region

Mashael M. Al Saud
Space Research Institute
King Abdulaziz City for Science
and Technology (KACST)
Riyadh, Saudi Arabia

ISBN 978-3-030-57630-1 ISBN 978-3-030-57631-8 (eBook)
https://doi.org/10.1007/978-3-030-57631-8



This Springer imprint is published by the registered company Springer Nature Switzerland AG
The registered company address is: Gewerbestrasse 11, 6330 Cham, Switzerland

# Preface

The Kingdom of Saudi Arabia, as the major part of the Arabian Peninsula, is known by its distinguished physical nature where the desert, as the most dominant landform, is interfered with mountain chains at different localities. However, this vast land area occupies tremendous unique natural and man-made features that make the Kingdom as a large museum.

In fact, the Kingdom of Saudi Arabia as well as many countries in the Arabian Gulf Region is known as oil-rich country, as well as with plenty of ore deposits. Thus, it is proposed that these natural resources are the reason behind the richness and wealth of the Kingdom; nevertheless, there are many factors contribute in the development of the country financial and social resources. There are frequent-wise leaderships of the Saudi government who are usually looking forward to raise the level of the Kingdom to the international levels and thus building developing projects and enhance the life quality and the income.

There is a vision by Saudi Government who noticed that the entire globe is witnessing geo-political conflicts and debate plus economic instability that lead to arguments between many countries. Thus, Saudi Arabia looks at the view from different side based on the development of the socioeconomic situation of the Kingdom apart from the income given from petroleum. It is really a Smart-Thinking and Recognition at a glance for optimal future.

Recently, NEOM Region has become a widespread term for the "Dream Zone," not only for the Saudi people, but also for people all around the world. It is not a matter of the distinguished geographic location with mild climate and landscape, it is a future strategic plan for national and international investments and the related activities. In fact, NEOM Region is an open hub for people from different races and regions who have also different levels of financial resources.

It is, therefore, a new innovative idea for gathering investments from one side and enhances the economical and social sectors of the Saudi Kingdom from the other side.

At the most northwest of the Saudi Kingdom, NEOM is a region with about 26,500 $km^2$ where it is located along the Red Sea and the Gulf of Aaqba and about 200 km faraway from Tabuk City. It is less than 15 km from Egypt and almost few kilometers from the border of Jordan. It has been announced by Saudi Crown Prince Mohammad bin Salman at the Future Investment Initiative conference in on October 24, 2017. Thus, Saudi Arabia started working in NEOM Region which is expected to be a junction between Silicon Valley, Dubai and Seychelles.

There are sixteen economic sectors, including water, energy, education, tourism, etc., determined to be the main drivers for NEOM future economy. Therefore, once NEOM attains an advanced stage of development, these sectors are anticipated to generate an estimated annual income of $100 billion.

As per the expertise of the author, who worked on several development topic for different regions from the Saudi Arabia, it was necessary to contribute by the author's knowledge and experience in supporting this project and provide analyzed data and information required. It is a private initiative taken by the author to participate in establishment of this global project.

This book will analyze and manipulate the natural and man-made characteristics of NEOM Region, and thus, it will recognize the existed natural resources and the threatening natural hazards in NEOM Region. This will help identifying the major aspects of the sustainable land management. In this respect, production of maps and the extraction of geo-spatial data and information were obtained by using advanced space techniques and the geo-information systems.

The author considers the production of this book as a national and ethical commitment in the series of her scientific productivity toward supporting the development programs and the empowerment of governmental institutions in the Kingdom of Saudi Arabia.

Riyadh, Saudi Arabia Mashael M. Al Saud

# Contents

# Acronyms

| | |
|---|---|
| AAII | American Association of Individual Investors |
| ADPC | Asian Disaster Preparedness Center |
| AGI | American Geological Institute |
| BRGM | Bureau de Recherches Géologiques et Minières |
| DEM | Digital Elevation Model |
| DHIS | Drill Hole Information System |
| DMMR | Deputy Ministry for Mineral Resources |
| EM-DAT | The International Disaster Database |
| FAO | Food and Agriculture Organization |
| GAMEP | The General Authority of Meteorology and Environmental Protection |
| GIS | Geographic Information System |
| INEGI | Instituto Nacional de Estadística y Geografía |
| IUCN | International Union for Conservation of Nature |
| LIF | Legatum Institute Foundation |
| MoA | Ministry of Agriculture |
| MODS | Mineral Occurrence Documentation System |
| MoEP | Minister of Economy and Planning |
| MoPMR | Ministry of Petroleum and Mineral Resources |
| NASA | National Aeronautics and Space Administration |
| NCWCD | National Commission for wildlife Conservation and Development |
| OECD | Organization for Economic Cooperation and Development |
| PPP | Public-private partnership |
| SCTH | Saudi Commission for Tourism and National Heritage |
| SDG | Sustainable Development Goals |
| SGS | Saudi Geological Survey |
| SLM | Sustainable Land Management |
| SWA | Saudi Wildfire Authority |
| UNDP | United Nations Development Programme |
| UNDPI | United Nations-Department of Public Information |
| UNEP | United Nations Environmental Programme |

| | |
|---|---|
| USEPA | United States Environmental Protection Agency |
| USGS | United States Geology Survey |
| WB | World Bank |
| WTO | World Tourism Organization |
| WWF | Worldwide Life |

# List of Figures

# List of Tables

# Chapter 1
# NEOM Region

**Abstract** On the upper part of north-west of the Kingdom of Saudi Arabia, NEOM Region is located along the Red Sea and the Gulf of Aaqba. This region has been designated, by the Saudi Government, to be an international commercial and industrials zone. This has been built on the available natural and unique components that are available in this region, added to the economic vision that Saudi Arabia sought. Thus, the idea of creating NEOM Region was viewed from different developing aspects including the national and international ones. NEOM Region, with about 26.500 $km^2$, is located in Tabuk Province and specifically at the most remote part of the Kingdom. If the spatial proximity of NEOM Region is considered, it is situated as a junction between three continents where it encompasses mountain ridges and plains, and thus it witnesses diverse climatic conditions which in turn result different biological components. The region of NEOM has several unique natural and attractive features, and it occupies considerable natural resources. The region has been remained undeveloped for many years where only dissipated human settlements and activities are located only along the coastal plain. It was just a pathway from the Kingdom of Saudi Arabia to Jordan and Egypt. This chapter will talk about the physiography of NEOM Region and then the existing natural and economical components upon which it has been selected, as well as it will discuss the development vision for the area as a global smart zone.

**Keywords** Coastal zone · Natural features · Biodiversity · Smart city · Development

## 1.1 Overview

The Kingdom of Saudi Arabia is the largest country in the Middle East and the 14th largest country in the World. It constitutes the largest part of the Arabian Peninsula (Al-Jazeera Al-Arabia) which is bounded by the Red Sea from the west, the Arabian Sea from the south and the Arabian Gulf from the east, and thus totaling about area of 2.15 million $km^2$. The Kingdom has a perimeter of about 6749 km where 4530 km is a terrestrial border and 2219 km shoreline. Thus, 1709 km extends along the Red Sea (including the Gulf of Aaqba) and 510 along the Arabian Gulf.



M. M. Al Saud, *Sustainable Land Management for NEOM Region*,
https://doi.org/10.1007/978-3-030-57631-8_1

Deserts occupy vast lands of the Kingdom where the Arabian Shield, including Najed Region, curvedly extends in the western side of the Peninsula forming Al-Hejaz and Aseer Mountain Ranges. The coastal plain along the Red Sea is named as Al-Tohama. While, An-Nafood Desert represents the northern part between Saudi Arabia, Iraq and Jordan. Besides, Al-Ruba'a Al-Khali Desert is located in the southeast of the Kingdom.

The population of Saudi Arabia is at about 35 million people in 2020 (Worldometer 2020), including about 5 million foreigners. The Kingdom is the birthplace of Islam where the Makkah Al-Mokarama and Al-Medina Al-Monawara are located and represent the holiest two Islamic cities where pilgrimage and Umrah are practiced.

The Arabian Peninsula in general and the Saudi Arabia in particular encompasses several environmental components, and thus it has distinguished biodiversity where unique flora and fauna are found. Hence, the natural interaction between mountains, plains, deserts and streams created remarkable geographic picture rarely match everywhere.

As the World's largest natural reservoir of petroleum, Saudi Arabia owns about 22% of the World's reserves of petroleum where the produced oil and gas is characterized by purity and shallowness to terrain surface, and then it can be exploited inexpensively. In addition, Saudi Arabia occupies several natural ores including mainly gold, phosphate, aluminum, copper, iron, etc. Saudi Arabia is classified into 13 provinces or regions (called Amara in Arabic) which are branched into 132 Governorates (Mohafazat) including 46 cities. Hence, Al-Sharqieh Province is the largest one (540,000 $km^2$), while the Al-Baha Province is the smallest (12,000 $km^2$), besides Riyadh District is about 380,000 $km^2$.

On the upper part of north-west of the Kingdom of Saudi Arabia, NEOM Region has been designated to be smart zone and incorporate function with economic and commercial cluster and tourist destination. It belongs to Tabuk Province (136,000 $km^2$) where it borders Egypt and Jordan, and at a range of 50 km from Tabuk City. This region is planned to cover a total area of 26,500 $km^2$ where the shoreline extends to about 225 km along the Red Sea (Fig. 1.1). Therefore, NEOM Region is situated between the following geo graphic coordinates:

27° 43' 27" N and 29° 08' 29" N and,

34° 31' 40" E and 35° 57' 55" E.

Therefore, NEOM Region is characterized by the following:

- It is almost a geographic junction between two continents, Asia and Africa, and between the Mediterranean Sea and Red Sea, where 10% of world trade flows through the Red Sea,
- It encompasses diverse geomorphology (e.g. altitude may exceed 2500 m), and then resulting typical biodiversity (i.e. unique flora and fauna) and diverse climatic conditions,
- NEOM Region occupies several archeological and heritage localities for different periods and aspects of civilizations,

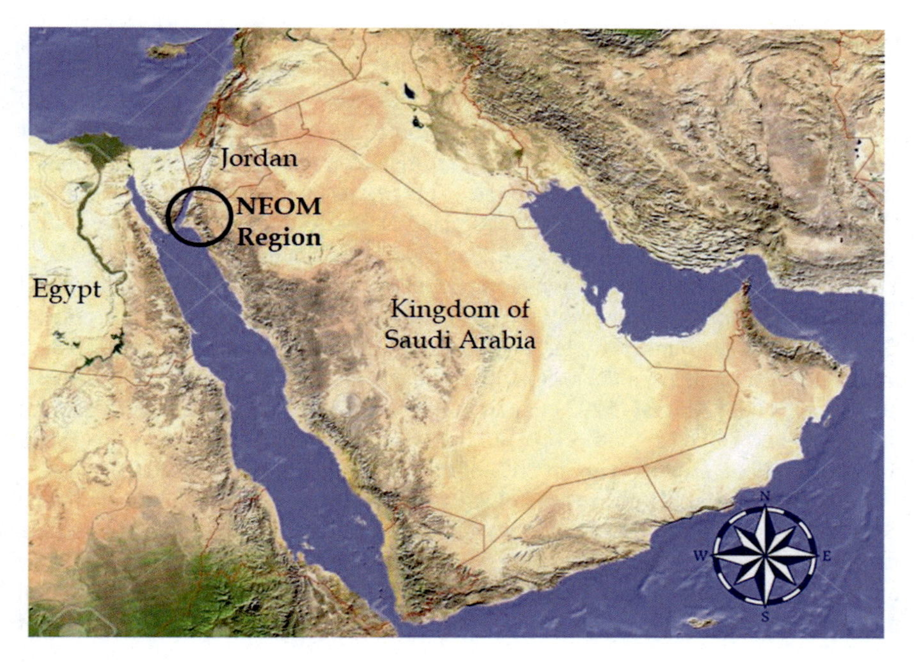


**Fig. 1.1** Location of NEOM Region

- It is rich with natural resources including mainly ore deposits, and many other potential resources,
- NEOM Region can be a centralized site in the upper most part of the Saudi Kingdom.

## 1.2 Western Area of Tabuk Province

While Tabuk Provenience is about 146,072 km$^2$; however, talking about western area of Tabuk Province points out to the location where NEOM Region has been designated. Even though NEOM Region occupies only 18% of Tabuk Province, yet NEOM Region distinguishes Tabuk Province with the natural characteristics that make it one of the most significant and remarkable proveniences of the entire Kingdom.

One of the most important characteristics of Tabuk Province is the long shoreline (648 km), which is about 38% of the western shoreline of Saudi Arabia (1709 km, including Gulf of Aaqba and Red Sea). Besides, NEOM Region constitutes about 40% of the northern shoreline of Tabuk Province.

NEOM Region encompasses few number of villages which do not exceed 25 villages with population of less than 1000 people in each. While there are 7 main towns, two of them (Al-Sharaf and Al-Bada'a) are located inside the region and the rest are coastal ones. The coastal towns are (from north to south): Maqna, Ayeynat, Al-Khrierbah, Gayal, Sharma, Al-Sorah and Al-Muwieleh.

Except Al-Bada'a, which is the only Mohafazat in NEOM Region with 13,000 people, and Sharma City with about 5000 people, the rest towns has population of less than 3000 people. Therefore, the total estimated population size in these villages and towns does not exceed 60,000 people. This means that the population density in NEOM Region is very low and it is estimated at about 2 person/km$^2$. This in turn reflects that the region of NEOM is still considered as a bare and uncovered area.

NEOM Region has three major geomorphological units, which will be discussed in details in Chap. 3. These are: the coastal zone, mountain ridges and the plateau (Fig. 1.2). It is surrounded by large towns along its border and these towns play a role in the development of the region. These towns are: Haqel (25,649 people) to the north-west, Halet Ammar (5293 people) to the north-east, Bir Hermaz (5975 people) to the east and Duba (25,568 people) to the south. The non-Saudi citizens in Tabuk and the mentioned towns is about 20% of the total population (GAS 2020).

Human activities in the region of NEOM are concentrated mainly along the coastal zone, while very few activities can be found in the mountainous areas where rugged topography is dominant. Therefore, coastal zone is characterize mainly by the touristic sites and resorts due to the remarkable sandy coasts where unique coral reefs can be observed. In addition fishing is a common job in the coastal zone of NEOM.

The agricultural activities in NEOM Region are located in the coastal zone and the flat plateau along the roads to Tabuk. Thus, palm trees, mango, orange and other fruits are found; in addition to different field crops and seasonal planting.

As a distinguished area, there are many archaeological and heritage sites that characterize in NEOM Region and the surrounding, such as Maghayer Chouaieb in Al-Bada'a and Bir Al-Sa'aidani in Maqna.

NEOM Region has remarkable transportation advantages. Thus, it constitutes the principal route between Jordan and other Saudi hubs where Al-Durrah border crossing is less than 50 km away from the border of NEOM Region. In addition,

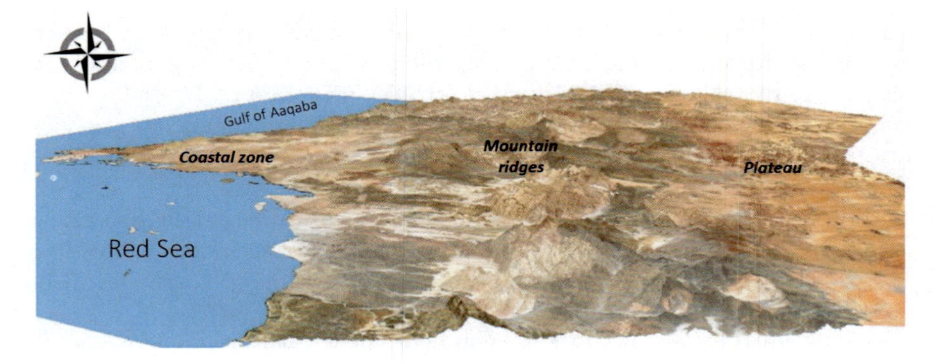


**Fig. 1.2** General space view (diagonal) for NEOM Region

Duba Port, as one of the major ports in Saudi Arabia, is about 15 km from southern border of NEOM border.

## 1.3 Vision for Development

If an inventory is applied; therefore, it can be reported that the Kingdom of Saudi Arabia occupies three major industrial zone which are among the top largest ranked of their type worldwide. These are: Al-Jubail Industrial City (920 $km^2$), Yanbua Industrial City (185 $km^2$) and KAEC Industrial City (181 $km^2$) where they are ranked respectively as 1st, 4th and 5th largest industrial zones over the entire globe (Worldatlas 2020).

It has been also reported that Saudi Arabia is ranked at 12th according to the International Standard Industrial Classification -ISIC, according to the WB and OECD (2018). While the kingdom is the 21st rank in tourism (WTO 2018)

According to Legatum Prosperity Index (LIF 2019), the Saudi Arabia, it ranked 71st of 167 countries. It showed that the Kingdom rises higher in prosperity ranking than any other MENA country. Thus, it has been increased from 54.75 in 2010 to 57.97 in 2019. Table 1.1 shows selected prosperity indices for Saudi Arabia.

According to Top Smart Cities Governments (TSCG 2019); however, none of the cities in Saudi Arabia is not included in the top 50 smart cities of the World where the indices used for ranking are tremendous including of vision, leadership, budget, financial incentives, support programmes, smart policies, etc.

Thus, it is sometime paradoxal that the Kingdom of Saudi Arabia would open this economical aspect and making a global commercial zone, while the Kingdom is known by its economical level where natural resources and most available. However, it is smart thinking to take this action.

**Table 1.1** Selected prosperity indices for Saudi Arabia

| Prosperity index[a] | World ranking | Prosperity index[aa] | World ranking |
|---|---|---|---|
| Market access and infrastructure | 46th | Entrepreneurship | 40th |
| Investment environment | 70th | Open business | 72nd |
| Governance | 68th | Power | 10th |
| Economic quality | 56th | Quality of life | 42th |
| Social capital | 42nd | Heritage | 50th |
| Education | 68th | | |
| Health | 68th | | |

[a]*According to Legatum Prosperity Index LIF (*2019*)*
[aa]*According to World Tourism Organization WTO (*2018*)*

### 1.3.1 What is NEOM?

The designation of NEOM is divided the word into two parts: the first includes the letters (NEO) which means in Greek the word "New", and the second letter (M) separately, and refers in Arabic to the word Mustakbal or Future in English; that is, the whole (NEOM) means the *"New Future"*. It is also referred to colloquially as "Noissa", in short for the English term "North West Saudi Arabia", which is an indication of its location in the northwest of the Kingdom.

In other words, it is a vision of what a New Future might look like in a typical geographic zone. It is a global initiative to make a civilized achievement different from what exist. It the time the World became in great need for fresh thinking and new measures for better life.

According to NEOM Brochure (2018), NEOM has been described as a living laboratory and hub for innovation. A sustainable ecosystem for living and working and a model for the new future. It will be also a destination, a home for people who dream big and want to be part of building a new model for sustainable living, working and prospering. The region is assigned to include towns, cities, transport networks, ports and enterprise zones, research centers, entertainment venues, and tourist resorts, etc. Eventually, it will be the home and workplace for more than a million citizens from around the world.

### 1.3.2 NEOM Vision

In October 24th 2017, the Saudi government took a decision to make NEOM Region as a centerpiece for Saudi Arabia's 2030 vision, and to promote the Saudi socio-economic status and enable the Kingdom to ride a pioneer role in the global development. Hence, Saudi Arabia aims to complete the first section of NEOM by 2025. The project has an estimated cost of $500 billion. Saudi Arabia aims to raise $100 billion.

Even though NEOM is a global project initially funded by the Kingdom of Saudi Arabia; however, the project will be achieved, incorporated and funded by foreign entities all over the World.

Therefore, NEOM will move consistently with sustainable development goals SDGs, and became be an international cluster where on new standards for community health, environmental protection and the effective and productive use of technology are set.

Therefore, NEOM is designed to be as (According to NEOM Brochure 2018):

1. An economic engine for the region and the World
2. A powerful approach to environmental conservation
3. A living laboratory where entrepreneurship will chart the course for a new future
4. A home for an international community of dreamers and doers
5. A site with breathtakingly diverse terrain and hospitable climate, coast, climate and desert.

### 1.3.3 *Motivations of the Study*

It can be a question: why to study "Sustainable Land Management" for NEOM Region? While the region is planned to be an economical hub; therefore, what is the interrelationship between the elements of land management and the economic aspects?

The author believes that for any developing project including the economic ones for any region, there must be a background knowledge on the natural and man-made influencers which act as criteria to characterize the region and its suitability to be an economic hub. Thus, for NEOM Region the following are the motivations to apply this study:

- Studies and investigations on NEOM Region are still rare, and perhaps focuses were on Tabuk Province as a whole while the western part of this provenience (recently designated as NEOM) has not given much attention. Therefore, the data available is not attributed to NEOM Region, but for the entire Province. For this reason, the identification of all data and information, belong to natural and economical components, must be a prior step. This helps determine the most suitable localities where projects of different purposes can be achieved.
- In case some thematic data are available for the western part of Tabuk Province, this was found as a comprehensive figure and not dedicated for NEOM Region itself. In other words, thematic maps specifically elaborated for NEOM Region should be available and then the thematic data must be assigned within the limited of NEOM.
- It is utmost significant to put data and information available on the natural resources as well as natural hazards for NEOM Region. This is essential for further investigation as major components of land management.
- NEOM Region is a vast land with rugged and remote localities in many instances. Therefore, the use of space techniques is a helpful and necessary tool. In particular, the use of satellite images will assure retrieving geo-spatial data and information which are difficult to be done by conventional methods. This is also applied for the use of Geographic Information System (GIS), which is an essential tool of geo-spatial data analysis, manipulation, mapping and storage.
- Except some head titles and broad objectives, there is no obvious plan put for land management for NEOM Region till quite recent. This is well noticed during several data screening and literature reviews done by the author on NEOM.

## 1.4 NEOM and SDG Vision 2030

The initiative behind execution of NEOM Region has been sought from a sustainable development vision for the next decades, and thus it meets with the global initiative of the Sustainable Development Goals (SDGs). These goals occupied wide attention for many custodians (e.g. UN-Water, UNESCO, FAO, UN-Environment,

WHO, UNECE, etc.) who are tackling with the developing issues on the global level. Therefore, the 17 designated sustainable development goals have been adopted by the United Nations in 2015 to work towards 2030.

The SDG framework provides a scheme for a sustainable future with goals traversing: poverty, health, education, water, climate change, forests, oceans and urban settlements (UNDPI 2012).

On the national level, the transformation needed to achieve the SDGs requires a united mechanism from all levels, where. The role of the Public-private partnership (PPP) and organizational SDG action is essential component of the agenda's implementation.

Thus, the Kingdom of Saudi Arabia is committed to implementing SDGs and gives the highest priority to this endeavour, as commensurate with the Kingdom's specific context and national principles (MoEP 2018). This has been assured by the Crown Prince of the Kingdom of Saudi Arabia Mohamad bin Salman who stated that: "In order to preserve the unique environmental character of the region, environmental sustainability laws and mechanisms will be developed. Natural resources will be conserved in accordance with the best practices and standards in place globally".

Implementing SDGs is consistent with the vision of NEOM Region development. Therefore, applying SDG would be helpful tool, and strategic endeavors for this global economic hub. However, the framework of SDGs applicability should be well illustrated.

In order to mainstream SDGs in the strategic vision of NEOM Region, the 17 goals should be matched with the objectives that NEOM Region has been created for. This can be demonstrated in Table 1.2.

## 1.5 Previous Studies

As mentioned previously in this chapter, that studies have been applied for NEOM Region in particular, because it is just recently designated region for development and its geographic extent has been lately determined. Thus, if studies have been carried out for this region, thus they would be done only in the last three years (since 2017) after NEOM project has been promoted.

However, there are many studies obtained on Tabuk Province and the surrounding where the recently designated NEOM Region is included. These studies were done on different topics, where few number of them can be useful supportive background knowledge for studying the sustainable land management of NEOM Region, and the largest number of these studies were used as references in this document.

In addition to the thematic maps (e.g. topography, geology, etc.) obtained for the entire Saudi Arabia, examples of these studies, which have been obtained on: geomorphology and soil (El Batawesy et al. 2013), climate change and its impacts (Al Zawad 2008; Al-Mutairi et al. 2019), hydrology (Al Saleh 2017), land cover/use (Al-Harbi 2010; Al-Balawi et al. 2018), floods (Abdelkarim et al. 2019) and natural hazards (Jones 2001; Theilen-Willige and Wenzel 2019).

**Table 1.2** Matching SDG and NEOM Region objectives

| SDGs | Major objective | NEOM objective |
|---|---|---|
| Goal 1 | No poverty | Sustainable production and secured food supply |
| Goal 2 | Zero hunger | |
| Goal 3 | Good health and well-being | Healthy and safe community |
| Goal 4 | Quality education | Educated society with literacy |
| Goal 5 | Gender equality | Embower women involvement |
| Goal 6 | Clean water and sanitation | Securing pure water supply |
| Goal 7 | Affordable and clean energy | Providing green energy |
| Goal 8 | Decent work and economic growth | Providing job opportunities and acceptable income |
| Goal 9 | Industry, innovation and infrastructure | Industrial and innovation zone par excellence |
| Goal 10 | Reduced inequality | Harmonizing society towards equality |
| Goal 11 | Sustainable cities and communities | Smart zone with sustainable visions |
| Goal 12 | Responsible consumption and production | Society with self-sufficiency |
| Goal 13 | Climate action | Adopting adaptation and mitigation measures |
| Goal 14 | Life below water | Reserving the marine systems |
| Goal 15 | Life on land | Reserving the terrestrial systems |
| Goal 16 | Peace and justice strong institutions | Global zone where people from different entities meet |
| Goal 17 | Partnerships to achieve the goal | NEOM will be a partner and pilot area for SDGs |

# References


Abdelkarim, A., Gaber, A., Youssef, A., & Prandhan, B. (2019). Flood hazard assessment of the urban area of Tabuk city, kingdom of Saudi Arabia by integrating spatial-based hydrologic and hydrodynamic modeling. *Sensors (Basel), 19*(5), 1024. https://doi.org/10.3390/s19051024.

Al Saleh, M. (2017). Natural springs in northwest Saudi Arabia. *Arabian Journal of Geosciences, 10*, Article number: 335.

Al Zawad, F. (2008). Impacts of climate change on water resources in Saudi Arabia. In *The 3rd International Conference on Water Resources and Arid Environments and the 1st Arab Water Forum.*

Al-Balawi, E., Dewan, A., & Corner, R. (2018). Spatio-temporal analysis of land use and land cover changes in arid region of Saudi Arabia. *International Journal of Geomate, 14*(44), 73–81. https://doi.org/10.21660/2018.44.3708.

Al-Harbi, K. (June, 2010). Monitoring of agricultural area trend in Tabuk region—Saudi Arabia using Landsat TM and SPOT data. *The Egyptian Journal of Remote Sensing and Space Science, 13*(1), 37–42.

Al-Mutairi, K., Alfifi, A., Aljahni, S., Albalaw, A. (April, 2019). Climate changes knowledge and awareness among people in Tabuk region, Saudi Arabia. *Acta Scientific Agriculture, 3*(4).

El Batawesy, M., Ramadan, R., Al Harbi, K., Fadi, A. (2013). Impact of the geomorphology and soil management on the development of waterlogging in closed drainage basins of Egypt and Saudi Arabia. *Environmental Earth Sciences, 68*, 1271–1283.
GAS (General Authority for statistics). (2020). The General Population and Housing Census 2010. Available at: https://www.stats.gov.sa/ar/14.
Jones, D. (2001). Blowing sand and dust hazard, Tabuk, Saudi Arabia. In J. S. Griffiths (ed.), *Book entitled: Land surface evaluation for engineering practices*. Geological Society Engineering Geology, special publication No. 18, p. 175.
LIF. (2019). The Legatum Prosperity Index. Available at: https://www.prosperity.com/globe/saudi-arabia.
MoEP (Minister of Economy and Planning). (9–18, July 2018). Towards Saudi Arabia's sustainable tomorrow. First Voluntary National Review 2018–1439, Kingdom of Saudi Arabia. UN High-Level Political Forum 2018. "Transformation towards sustainable and resilient societies", New York. p. 166.
NEOM Brochure. (2018). The story of NEOM. Available at: https://www.neom.com/en-us/.
Theilen-Willige, B., Wenzel, H. (2019). Remote sensing and GIS contribution to a natural hazard database in Western Saudi Arabia. *Geosciences, 9*(9), 380. https://doi.org/10.3390/geosciences9090380.
TSCG. (2019). Top 50 Smart Cities Governments 2019. Available at: https://www.smartcitygovt.com/.
UN Department of Public Information. (2012). *UN General Assembly's Open Working Group Proposes Sustainable Development Goals*. United Nations: New York, USA.
WB and OECD. (2018). World Bank national accounts data, and OECD National Accounts data files. Available at: https://www.indexmundi.com/facts/indicators/NV.IND.TOTL.CD/rankings.
Worldatlas. (2020). The World's Largest Industrial Areas. Available at: https://www.worldatlas.com/articles/world-s-largest-industrial-areas.html.
Worldometer. (2020). Saudi Arabia population. Available at: https://www.worldometers.info/world-population/saudi-arabia-population/.
WTO. (2018). Yearbook of Tourism Statistics, Compendium of Tourism Statistics and Data Files. Available at: https://www.indexmundi.com/facts/indicators/ST.INT.ARVL/rankings.

# Chapter 2
# Land Management

**Abstract** Land management is usually performed within the context of national plans and strategies in order to put the frameworks of planned hubs. Thus, the selected themes for studying land management are induced by dimensional aspects of the study area. Hence sustainable land management (SLM) became a road map and preparatory phase before applying any project whatever its themes are, such as themes on economic, urban, agriculture etc. Even that, the success of many executed projects worldwide has been depended on the creditable and successful SLM studies done for these projects. Therefore, studies on land management have been adopted in most of the projects executed in the Kingdom of Saudi Arabia, and they became conditional terms of reference in these projects. However, SLM studies follow different methods and they use miscellaneous of tools. Therefore, the selection of methods and tools are significant to attain with successful results. NEOM Region, with its geography and natural resources as well as the physical processes occur on its land, is a typical and perspective zone where SLM should be carried out. However, much precision compared with other regions should be given to NEOM Region, because it will be a global hub. This chapter discusses the principles of SLM and correlate these principles with the constraints and implementations of NEOM Region, and its applicability to resources management and natural hazards management; eventually, the chapter will show the framework of the applied study in this book.

**Keywords** Environment · Urban expansion · Economic zone · Natural variables · Coastal zone

## 2.1 Concepts

In accordance with the principles and framework of land management, natural resources and lands can be used with beneficial manner where appropriate land management methods are applied in the context of regional planning and development strategies. If sustainability factor is considered; therefore, it will be described as sustainable land management (SLM).



M. M. Al Saud, *Sustainable Land Management for NEOM Region*,
https://doi.org/10.1007/978-3-030-57631-8_2

### 2.1.1 Definitions

There are many definitions for SLM, where most of them are contradicted with each other, and some of them focus on specific themes. In this respect, Smyth and Dumanski (1993) SLM is achieved in order to harmonize objectives of providing environmental, socio-economic, opportunities for the current and future benefits, while maintaining the quality of the land resource (soil, water and air).

UNECE (1996) defines land management as the process by which managing land resources are put to good effect, and with the most optimal use. It includes all implementations concerned with the management of land as a resource from an environmental and economic perspective. Thus, it can include farming, mineral extraction, property and estate management, and urban planning.

Another, and much more indicative, definition for SLM has been illustrated by TerrAfrica (2005) where SLM was described as: the adoption of land-use systems that through appropriate management practices enable land users to maximize the economic and social benefits from the land while maintaining or enhancing the ecological support functions of the land resources.

The World Bank describes SLM as the knowledge-based method that helps integrate several components including land, water, biodiversity, and environmental management (including the existing challenges) to meet the increased food demands while insuring the sustainability of ecosystem services and livelihoods (WB 2006a).

Moreover, World Bank identifies SLM as a process in a charged environment between environmental protection and the guarantee claim of ecosystem services on the one hand. On the other hand, it is about productivity of agriculture and forestry with respect to demographic growth and increasing pressure in land use (WB 2006b).

Mitiku et al. (2006) stated that SLM can be approached by looking at the symptoms of unsustainability which may involve land degradation and soil erosion, deterioration in water quality, degradation of biodiversity, occurrence of plant insect and diseases, etc.,

Liniger et al. (2011) stated that SLM is the increasing of land productivity and this entails reducing high water loss through run-off and unperceived evaporation from unprotected soil, improve infiltration. It also includes improved livelihoods and improved ecosystems: being environmentally friendly.

There are numerous technologies and practices to achieve SLM, and these are usually used to address land degradation which affects about one-third of global terrestrial area and is having negative impacts on the incomes and food security of agricultural populations (Mirzabaev 2016).

The Food and Agriculture Organization (FAO 2020) defines SLM as the use of land resources, including water, soils, livestock and vegetation for the production of goods to meet changing human demand, while simultaneously ensuring the long-term productive potential of these resources and the maintenance of their environmental functions.

Therefore, SLM can be simply described as a tool for managing Earth's resources and land components from a sustainable point of view. Thus, improper land management can lead to land degradation and distinctive reduction in the functions of productivity and services.

Form the mentioned examples, it was noticed that the adopted definitions for SLM that focuses are always on three pillars. These are:

- Natural resources and ecosystems (including largely, soil, water, livestock and ecologies),
- Food security and economics and production of goods,
- Land and resources degradation (soil erosion, water quality deterioration, etc.).

Therefore, it worth mentioning that SLM is neither adopted for the trend- analysis of resources and food and agricultural security nor for the assessment of natural processes. Nevertheless, SLM must be always viewed as planning and frameworks build up schemes.

Alternatively, if SLM would be assigned for a specific theme; therefore, it must be named accordingly. For example, it can be named as "sustainable agricultural land management" for agronomical managing and planning and so on.

### 2.1.2 Objectives of SLM

Sustainable land management is carried to reach an optimal management, use and development of land resources which are utilized for many purposes including (but not limited to) agriculture, water resource, urban planning, eco-tourism, etc. Thus, in the achievement of SLM, the existing natural and man-made components, as well as the potential changes must be accounted. In addition, implementing derivers should be assessed.

In this view, SLM objectives are always found different from one project or study to another. However, the most adopted ones are accorded with those put by Smyth and Dumanski (1993). These are:

- Enhance and maintain productivity,
- Reducing the level of production risk, and promote balance between a land use and prevailing environmental conditions (stability/resilience),
- Protect the potential of natural resources and prevent degradation of soil and water quality,
- Maintaining economic viability otherwise land uses will not sustain,
- Accepting the social impact and assure access to the benefits from improved land management (acceptability/equity) where the populations most directly affected by social and economic impact.

Therefore, it is usually tedious to identify the exact objectives of SLM unless all acting factors are determined or the outline of SLM is adopted for specific themes.

Hence, the most important factors which should be accounted while applying SLM are:

1. Project scale, which is significant to determine the dimensional aspects of any project where SLM would be applied to. Thus, small-scale projects do not have similar SLM method as that of large-scale ones.
2. Available financial resources are utmost important and they are always joined with the project dimensions,
3. Priority themes, which are the type and number of themes (e.g. agriculture, tourism, etc.) to be investigated in the context of SLM,
4. The natural conditions and influencers of the area of study is significant and must be assessed before SLM method is applied in order to put the influencing factors in the work scheme.

In accordance with the requirements for NEOM Region, the author can describe the objectives on applying SLM study for this region as follows:

1. NEOM Region is still unstudied and uncovered geographic zone, and making it a hub for smart cities with global economic centres, needs to have initially the characteristics of its land identified and organized for further works needed,
2. NEOM Region is well known by its diverse geomorphology, where the latter controls many other natural components, such as climate, topography, etc. This in turn requires in-depth data analysis for the diverse land characteristics which can be attained by SLM methods,
3. The objectives behind establishing NEOM Region are wide enough and tackle several features and aspects which compel identifying a number of land components and their influence on the management approaches needed,
4. There is no unified and comprehensive physical data (e.g. geology, geomorphology, climate, landforms, etc.) investigation for NEOM Region. Therefore, SLM will put the detailed inventory for all these data to the fact,
5. Neither natural resources nor natural hazards have been studies for NEOM Region, thus the optimal SLM will tackle all aspects of these two pillars in details, where they will introduce the incentives and zones under risk in one package in order to harmonize the selection of the suitability of locations for different purposes,
6. Up to date, there is no SLM study has been achieved yet for NEOM Region which must be an initial and principal action taken for such a giant project.

## 2.2 SLM-Related Constraints

Land is a limited resource component, and hence the economic competition is being escalated as a result of the rapid and increased urbanization, exacerbation in population rate, socioeconomic needs and the increased demand for resources.

Therefore, setting land management systems are institutional frameworks intricate by the actions they must perform, by national cultural, political and legal settings and by technology (Enemark 2005). In addition, countries should set up land management systems considering responsibilities and the existing constraints that may create in managing the relationship between land and people.

The achievement of SLM is usually faced with several constraints occur whether before or during the elaboration of SLM. Hence, it would be significant to highlight these constraints at the early stage of establishing SLM systems. Therefore, the most known challenges are:

1. Natural constraints: In many instance, natural features can be as constrains due to many unfavorable factors, such as rugged topography, inaccessible areas and dangerous sites,
2. Data lacking: If thematic data, records and information are not available, it would be a problem to apply SLM, unless new works would be carried out to generate database, and this costs time and resources,
3. Expertise: Recently, advanced techniques and sufficient knowledge are significant to elaborate SLM, but when they are not available it would be difficult to make creditable SLM,
4. Financial resources: Expenses are required to achieve optimal SLM and the lack of financial resources often represent a problem,
5. PPP (public-private-partnership): The coordination between the public and private sectors is utmost significant, but usually it does not exist and troubles occur during several stage of work,
6. Integration with stakeholders: There is usually weak contact between inhabitants (where SLM to be applied) and rest stakeholders including mainly decision makers. This often results constraints in work achievement,
7. Property rights: It is a frequent problem occurs when lands included in land management are owned by the individuals. This is well pronounced and often makes problems,
8. Environmental controls: This includes wide aspects of environmentally-related features. Examples of these feature (but not limited): waste disposal sites, forest fragmentation, air pollution due to industrial activities, etc.,
9. International markets: The local and international markets are changing and this results economic oscillations that may retard the optimal economic cycle where SLM will be applied,
10. Geopolitical constraints: In many instances problems occur when the geography of a land is shared between two or more countries, such as in the case of shared water resources, undefined borders, etc.

These constraints can exist if SLM is adopted for NEOM Region, but they will be at different levels as shown in Table 2.1.

**Table 2.1** SLM-related constraints and their existence in NEOM Region

| SLM-related constraints | Existence in NEOM Region[a] |
|---|---|
| Natural constraints | There is rugged topography, inaccessible areas |
| Data lacking | Some data available but it needs compiling, other data needs to be elaborated |
| Expertise | Almost available |
| Financial resources | Financial resources are almost secured |
| PPP | PPP is weak |
| Integration with stakeholders | Must be reinforced |
| Property rights | Problems of property rights exist |
| Environmental controls | Should be considered as a major theme in SLM |
| International markets | Future economic scenarios should be adopted |
| Geopolitical constraints | Totally solved (with Egypt) |

[a]*Existence before or during SLM application*

## 2.3 Resources Management

Resource management is the mechanism of handling all existing resources effectively. It includes planning and then all resources are assigned to define tasks. Thus, applying proper resources management needs to prepare financial resources, projects and plans, equipment, and expertise.

In resources management, usually the available resources must be determined first, and this includes the assessment of their bulk and feasibility of investment in order to deduce their benefit and sustainability. In this respect, the major goal of resources management is to provide methods with the related guidelines and plans to facilitate the management of project people resources.

### *2.3.1 Implementing Resources Management*

For optimal resources management, there are four pillars to be considered in order to achieve a successful project that meets with as socioeconomic benefits on the national level. These are as shown in Fig. 2.1:

1. Resources identification: It is the initial phase when resources are recognized whether spontaneously or by applying the scientific methods. This represents the raise of primary idea on resource investment for the enhancement of the socioeconomic status. In this respect, the dimensional aspects (e.g. geographic extent, reserves, exposures, etc.) are accounted and then promoting the idea for the development will be followed.
2. Investment visibility: There can be resources available, and sometimes even with considerable extent, but their investment will result little output due to many

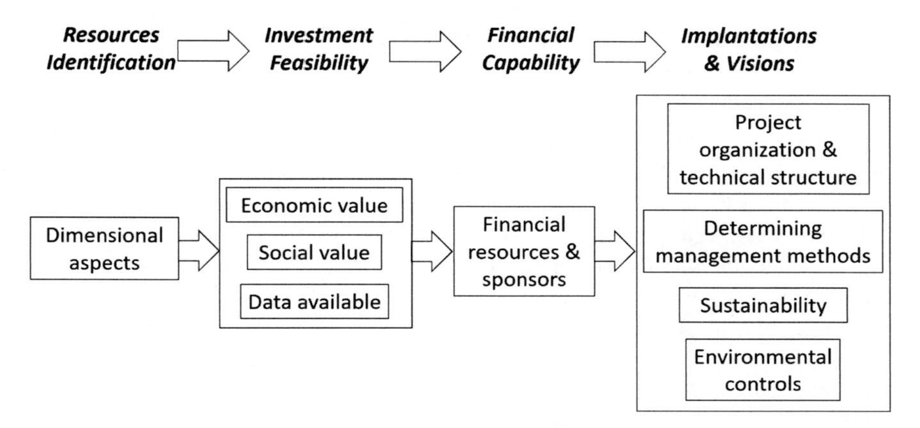


**Fig. 2.1** Major pillars and their associated items for optimal resources management approach

reasons including mainly the limited size of resources, difficulty of exploitation or high expenses required. Therefore, the economic value should be considered as one of the major targets.
Visibility dimensions do not limited for the economic value, but the social impact is also an output. In this respect, the optimal investment will open the chance for employment opportunities, and it results demographic changes.
In addition, visibility of investment is also viewed from the data availability which is a main component in resources management approaches.

3. Financial capability: As a principal factor in implementing resources management, financial resources should be available and the balance between investment cost and output must be primarily calculated. Therefore, when financial resources are not available, there will be difficulty, and sometimes stop proceeding, in taking actions forward for the investment. In this respect, there can be sponsoring of financial resources that can be adopted by combination between different investors where foreign investors can be involved.
4. Implementations and visions: When the first three pillars are adopted; therefore, decision can be taken for the implementation and this requires several preparatory and organizing endeavors including project organization (i.e. personnel and technically) where tasks per mangers, experts, technicians and labors are determined. This also include the technical methods to be followed in the investment approaches where scheduling and timeframe are fixed.

Other than the direct implementation procedures and preparedness approaches, there is always a vision considered to deduce the sustainability of workability and productivity of resources. In addition, the environmental controls (e.g. avoiding pollution, conserving landscape, etc.) are usually accounted where the environmental limitations are respected.

### 2.3.2 Resources Management for NEOM Region

Studies on resources management in Saudi Arabia are few or sometimes they tackle general discussion, such as Thompson (2018), but they can be found on specific themes, such as those on water resources and urban planning (Mohoji 2007, FSC 2017). Whereas the issue of resources management in the Kingdom are almost done on the individual basis. For example, local inhabitants of the southwestern part Saudi Arabia appear to have managed successfully to balance population growth with natural resources under a local tribal self-government (Eben Saleh 1998). But this is not always the case for many other parts of the kingdom of Saudi Arabia.

Neom Region, which is still a bare land and has been witnessing only slow population and economic growth, is a typical example where no resources management approaches have been applied yet, but the initiative, taken by the author, of establishing a global economic zone raised the issue and draw attention to SLM in general including the existing resources there. This has been dependent on the thematic maps available; especially the geologic, topographic and land form maps, in addition to the field reconnaissance which was carried out by the author in NEOM Region.

If the associated items of the identified pillars for applying resources management approach, which was described in Fig. 2.1, are allocated to NEOM Region; therefore, different levels of identification can be found as in Fig. 2.2. However, the overall figure shows that there is still lacking in identifying all items required to manage resources in NEOM Region.

Figure 2.2 shows that the presence of resources, data and financial resources are almost identified, and this identification of resources presence are touched by different stakeholders, whether by individual, experts or even the decision maker. While, data is not well identified, except that thematic maps are present but they are not dedicated for NEOM Region, but for the area of Tabuk and the surrounding. Besides, the financial resources are allocated for establishing NEOM Region and this was disseminated via media.

There are also some associated items which are slightly identified for the resources management in NEOM Region as shown in Fig. 2.2. These are the economic value of applying resources management approaches, as well as project organization and its sustainability. Whereas, social values (or impact), determining methods for the

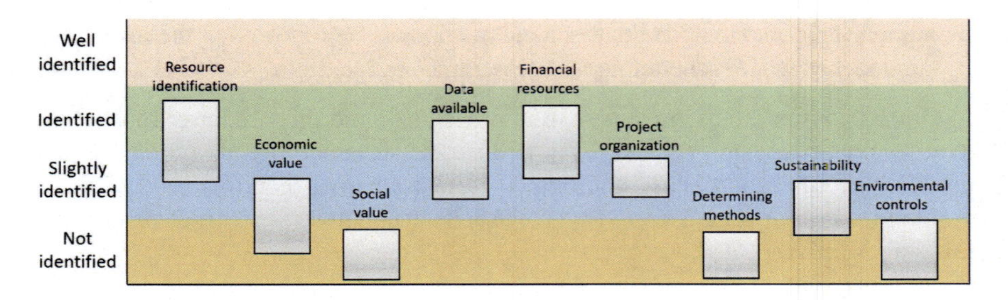


**Fig. 2.2** Presumed levels of identification for resources management in NEOM Region

management and environmental controls are nor identified yet, or at least they are not known by individuals.

There are some traditional and old methods to manage resources on a small-scale in NEOM Region; however, new integrated programs can combine scientific knowledge and practice with traditional methods. Nevertheless, no resources management plan for NEOM Region has been revealed up to date, and this motivated the author to carry on identification of the main resources located in the region and more certainly the natural resources which can beneficially invested in the context of the establishment of the global economic zone (NEOM Region).

For NEOM Region, two major aspects of natural resources have been tackled by the author. These two resources, which are the groundwater and ore deposits, are already under exploitation, but with no investment plans or even organization in the way of investment.

It is worth mentioning that more potential natural resources are available in NEOM Region, more specifically petroleum, but detailed studies and prospecting are required to assure their existence that is why this author did not include them in the SLM for NEOM Region.

## 2.4 Risk Management

Risk management has been followed since long time in the western part of Tabuk Province, the current NEOM Region, where small-scale and traditional methods for controlling natural hazards were applied. Thus, any breakdown of these traditional methods of risk management now prevents indigenous residents of the area from protecting the human and environment of the region.

The rugged morphology in combination with the complicated geology and lately the interference caused by the changing climate results unstable natural conditions that are represented by aspects of natural hazards in NEOM Region. Therefore, this geo-graphic part, likewise most of Arabian Peninsula area, is under natural risk and this should be taken into account while establishing SLM in the view of the execution of the global economic zone.

There are three types of natural hazards occur in NEOM Region. These hazards mangle between geomorphological, geological and climatic elements. Therefore, the majority of these hazards implies floods, terrain instability and seismic activities, which are common risky phenomena in the entire Kingdom of Saudi Arabia.

From the land management point of view, these aspects of natural hazards should be considered and they must be as an initial phase for SLM. Thus, the possible geographic distribution of these hazards must be recognized as well as the potential impact level, notably areas with hot spots, should be determined. This would be done in order to avoid executing projects and works in the areas under risk as well as to take appropriate controls. Thus, the following points were viewed while studying the natural hazards in NEOM Region:

1. The already existed natural hazards in the area of study, and all available information and historical records (if exist) about these hazards, with a special focus on their impact,
2. Location of the existing and planned urban activities and projects (human settlements, industrial zones, etc.) with respect to the hazardous areas,
3. Possibility of avoiding the hazardous sites or making shift to another safe sites,
4. Investigating the located controls, whether on individual or public level, to reduce or mitigate the impact of these risks,
5. Identifying the natural and man-made factors act in creating these hazards.

## 2.5 Framework of the Study

The idea behind preparing this document is to assure that all required data for applying this global project are available and if not they must be prepared accordingly. It also aims at introducing background knowledge about the natural setting of the region, and the probable development schemes. Therefore, this document can be a first-hand information for further and in-depth investigations needed to elaborate any theme representing a major part of the project achievement.

There are five main phases put in this document to compose the entire sustainable land management elements. They include naturally occurred elements, and the integrated ones by human activities, as well as the proposed management components. In addition, positive and negative influences are determined. This includes the existing natural resources, as positive influencers, as well as the natural hazards as negative ones.

Thus, the five phases, as shown in Fig. 2.3, are:

1. Database: It is the preparation of all data and information required to establish database on the area of study. They can be use also to elaborate thematic maps and data analysis. These datasets include climatic and hydrologic time series, seismic records, available maps and variety of socioeconomic data.
2. Thematic maps: The geo-spatial distribution of different terrain components should be accounted while setting the land management approaches. These are maps for all natural components which are needed for studying the physical setting of the region and the processes occur there. This is including: topography, geology, drainage systems, slope, landforms, and Digital Elevation Model (DEM).
3. Natural resources: There are several examples of natural-resource-based development, and thus it is well known that natural resources are endogenous to the economy and can develop important dynamic linkages (Andersen 2012). Therefore, natural resources are sort of incentives for working in any region. Thus, the available natural resources that can be used in the development of the area of concern were investigated. In this respect, raw materials and ores in NEOM Region were identified, as well as the groundwater resources were also studied.

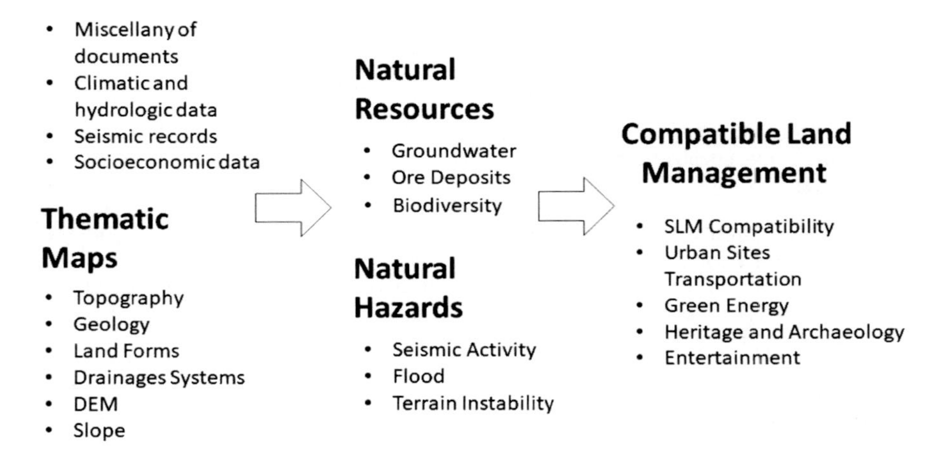


**Fig. 2.3** Phases for studying sustainable land management for NEOM Region

4. Natural hazards: The natural phenomena of an extreme nature, or natural hazards, has assumed increased importance as the population of the world and the value of the urban and economic investments at risk have grown (Oliver 1975). Therefore, identifying zones under natural risk is a must in order to avoid these zone from disasters and unfavourable conditions, and sometimes appropriate controls are applied. This document includes the assessment and mapping of the major natural risks. These are the flood, earthquakes and terrain instability.
5. Urban/Land planning: To achieve the prosperity in urban development, it is necessary to consider the urban planning in development, or the sustainable urban development (UN-Habitat 2013). Therefore, without obvious planning approaches to allocate activities and works in their appropriate locations. This documents tackled the major components of urban planning as main framework of the sustainable land management. This includes: urban settlements, green energy, touristic and entertainment, transportation and biodiversity.

## References


Andersen, A. (2012). Towards a new approach to natural resources and development: The role of learning, innovation and linkage dynamics. *International Journal of Technological Learning Innovation and Development, 5*(3), 291–324.

Eben Saleh, M. (1998). Planning for conservation: The management of vernacular landscape in Asir region Southwestern Saudi Arabia. *Human Organization, 57*(2), 171–180.

Enemark, S. (2005). Understanding the Land Management Paradigm", FIG Commission 7 Symposium Innovative Technologies for Land Administration, 19–25 June 2005, Madison–Wisconsin, USA.

FAO (Food and Agriculture Organization). (2020). Sustainable Land Management. Available at: http://www.fao.org/land-water/land/sustainable-land-management/en/.

FSC (Future Saudi Cities). (2017). Governance of planning local planning urban management programme. Department of Geography and Planning, University of Liverpool, p. 315.

Habitat, U. N. (2013). *State of the World's Cities 2012/2013: Prosperity of Cities* (p. 2013). Rout ledge: New York, USA.

Liniger, H., Mekdaschi, R., Hauert, C., & Gurtner, M. (2011). Sustainable Land Management in Practice—Guidelines and Best Practices for Sub-Sahara Africa. Technical Report published by TerrAfrica, p. 54.

Mirzabaev, A. (2016). Land degradation and sustainable land management innovations in central Asia. In book: Technological and Institutional Innovations for Marginalized Smallholders in Agricultural Development. DOI: http://doi.org/10.1007/978-3-319-25718-1_13.

Mitiku, H., Herweg, K., & Stillhardt, B. (2006). Sustainable land management: A new approach to soil and water conservation in Ethiopia. Mekelle, Ethiopia. ESAPP. p. 305. Available at: https://doi.org/10.7892/boris.19217.

Mohorji, A. (2007). Water resources management in Saudi Arabia and water reuse. *International Journal of Environmental Studies, 58*(5).

Oliver, J. (April, 1975). The significance of natural hazards in a developing area: A case study from North Queensland. geography. Published by: *Geographical Association, 60*(2), 99–110.

Smyth, A. J., & Dumanski, J. (1993). FESLM: An international framework for evaluating sustainable land management. A discussion paper. World Soil Resources Report 73. Food and Agriculture Organization, Rome, Italy, p. 74.

TerrAfrica. (2005). Improving country investment programming through advocacy, alliances, and alignment. In UN expert group meeting on sustainable land management and agricultural practices in Africa: Bridging the gap between research and farmers (April 16–17, 2009). University of Gothenburg, Sweden.

Thompson, M. (2018). Saudi Arabia: Civil society and natural resource management. In Overland, I. (ed.), *Public Brainpower*. Palgrave Macmillan, Cham, pp. 291–309. DOI https://doi.org/10.1007/978-3-319-60627-9_16.

UNECE (Economic Commission for Europe). (1996). Land administration guideline with special reference to countries in transition. Geneva, ECE/HBP/96.

WB (World Bank) (2006a). Sustainable Land Management. Challenges, Opportunities, and Trade-offs. Washington, DC.

WB (World Bank). (2006b). Sustainable Land Management. Washington, DC: World Bank.

# Chapter 3
# Physical Characteristics

**Abstract** Land management is a procedure to develop terrain surface and the existing components on it. Therefore, land management is integrally associated with the physical characteristics which control the existing components. Therefore, managing lands requires a detailed understanding on the acting natural influencers. In this respect, many elaborated studies based initially on the investigating of the major physical variables and on data available belong to these variables, which are also described as "physical characteristics". Thus, the acting variables become the pillars upon which land management approaches are based, and without identifying these variables, the applied approaches and the resulted outcomes will be uncreditable. For NEOM Region, no studies have been done yet to investigate the physical characteristics of the area, and there are only few thematic maps obtained for Tabuk Province within the context of mapping the entire Kingdom of Saudi Arabia. NEOM Region, with its diverse and complicated physical components, requires in-depth analysis for the existing physical characteristics. This is essentail becasue three major SLM pillars, including the management of resources, hazards and planning, are based on knowing these characteristics. This chapter will present detailed discussion on the geomorphology, climate, hydrology, geology and landforms of NEOM Region where this discussion will be oriented towards building the bases for further analysis of SLM applicability.

**Keywords** Mountain ridges · Precipitation · Rock types · Faults · Alluvial deposits · Red sea

## 3.1 Geomorphology

The geographic location of NEOM Region is considered as a major criterion that characterized it with good suitability to be a global economic zone. Therefore, NEOM Region has diverse geographic features and encompasses: coastal zone with different coastal plains, mountain ridges, crests and plateau. It is bordered by the Gulf of Aaqba from west and the Red Sea from south-western part, Al-Muwieleh and Jabal Al-Deba'a from south-eastern, Tabuk City and the surrounding from east, and Jordan from north.


M. M. Al Saud, *Sustainable Land Management for NEOM Region*,
https://doi.org/10.1007/978-3-030-57631-8_3

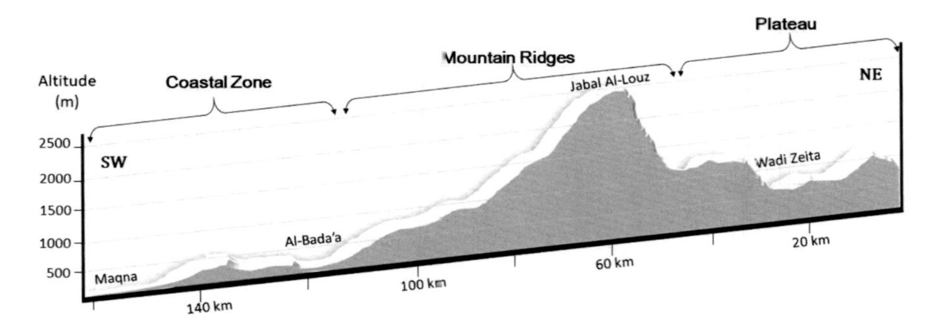


**Fig. 3.1** Simplified geomorphologic section for NEOM Region

According to several geomorphologic features, NEOM Region can be classified into three principal geomorphologic units. These are the coastal zone, mountain ridges and the plateau (Fig. 3.1).

### *3.1.1 Coastal Zone*

The shoreline of NEOM Region extends along the Red Sea where the largest part of the shoreline is adjacent to the Gulf of Aaqba. Thus, the total length of NEOM costal stretch is about 225 km where it encompasses several coastal portions which are controlled mainly by structural and geomorphological features. The estimated total area of the coastal zone is 8125 km$^2$ (31% of NEOM Region).

The most typical of these portions are located at (from south to north): Al-Sorah, Sharma, Al-Dniebah, Om-Ross, El-Khabat, Ras -Kasba, Ras El-Shyiekh Hamid, Sharam El-Mjawa and Ras Klieb.

The southern part of NEOM shoreline is characterized by geomorphological irregularity which has been resulted due to the impact between littoral and terrestrial processes including floods, erosion and sedimentation. Therefore, abruptly curved and distorted coastal lines are tremendous, but this is little pronounced in the western coastline along the Gulf of Aaqba.

The irregularity of the coastline, along the southern part of NEOM Region, resulted many islands and semi-islands with different shapes and dimension. Typical of these islands are (from south to north): Om-Qsoor, Al-Farsheh, Al-Thaghba'a, Sandeleh, Sanafeer and Thiran where the last two islands are the largest ones with an area of about 32 and 81 km$^2$; respectively.

The distance between the closest portion of NEOM Region and the Egyptian border is about 13 km at Ras El-Shyiekh Hamid, which is also the narrowest width of the Gulf of Aaqba, while the widest one is about 26 km at Maqna.

The coastal zone of NEOM Region constitutes the coastal plains mountain footslopes. Thus, they can be described as follows:

1. Coastal plains—Most of coastal plains are located mainly in the south-western part of NEOM Region where they largely compose talus slopes and alluvial fans for miscellany of deposits (e.g. alluvial and colluvial deposits mixed with marine sands and dunes) derived along the existing streams. Therefore, except one coastal plain located along the Gulf of Aaqba, all other plains show V-like shapes and connected with valley courses. The main coastal plains are:
   - Al-Muwieleh-Al-Sorah (up to 4, 32 km length),
   - Sharma-Al-Khrierbah (up to 10.5, 25 km length),
   - Gayal (up to 3 and 15 km length),
   - Al-Khabath-Ayeynat (up to 36 and 42 km length). This plain with Gayal plain are described by Clark (1987) as Lisan Basin,
   - Ras Klieb-Al-Hamida (up to 11 and 51 km length).
2. Mountain foot-slopes—These are the gentle slopes and hill adjacent to the coastal plains of the NEOM Region. They are characterized by diverse slope degree and altitudes, where the average slope gradient ranges between 10 and 40 m/km, while the average altitude ranges between 40 and 120 m above sea level. Hence, they can be classified as mountain hills that situated as follows:
   - At a range from the shoreline (i.e. several tens of kilometers). These are: Jabal Sharah, Jabal Khalsa, Jabal Hareb, Jabal Al-Shayati and Jabal Jbar).
   - Directly on the coastline and sometime with minimal distance from it (<500 m). They are mainly along the shoreline of the Gulf of Aaqba. These are: Jabal Al-Rughamah, Al-Radah, Jabal Al-Risham and Jabal Zuriek.

### *3.1.2 Mountain Ridges*

Mountain ridges are those mountain chains extending from the hills of the coastal zone upward to the adjacent mountains of NEOM Region (Fig. 3.1). They are sometimes forming the extension of the mountain hills to the east and north. Thus, the mountain ridges encompass an area of about 12,720 km$^2$ (48% of NEOM Region).

Mountain ridges can be classified into mountain hills and crests as follows:

1. Mountain hills—Mountains constitute the main body of NEOM Region where they extend in the NW-SE direction and separate the coastal zone form the plateau. They are characterized by different slope gradients, but they are in general represent rugged topography with steep slope where the slope gradient exceeds 120 m/km in many instances. While the average altitude ranges between 800 and 1400 m above sea level.
2. Crests—These are the highest mountains in NEOM Region where they are distributed within the mountain ridges. These crests, which can be observed from several remote areas, represent the tops of the most abrupt mountains in the entire region. They are almost found with altitudes exceeding 1600 m.

The most known crest are (from south to north): Jabal Al-Daher (1626 m), Jabal Terban (1612 m), Jabal Al-Shayati (2103 m), Jabal Al-Samakh (1762 m), Jabal Al-Da'ba'a (2316 m), Jabal Zuhed (1980 m), Jabal Houdh (1832 m), Jabal Al-Louz (2401 m), Jabal Fayhan (2549 m), Jabal Al-Kloom (2398 m), Jabal Al-Na'ayjat (1818 m), Jabal Amiq (1766 m) and Jabal Thoughob (1889 m).

### *3.1.3 Plateau*

The plateau of NEOM Region, rarely mentioned before, is the area which represents the flat land that extends from the mountain ridges towards Bir Hermaz and then Tabuk to the east. It is described by Clark (1987) as Hadabat Hisma. It has an area of about 5565 km$^2$ (21% of NEOM Region). Thus, the plateau is gently sloping to the east with as slope gradient less than 10 m/km. While, the altitude is almost between 900 and 700 m from west to east.

The plateau is shaped from the west (i.e. at boundary with the mountain ridges) by sharp shelf that extends in the NW-SE direction (Fig. 3.1). Thus, the geomorphology is totally changed from this shelf to the east (all along the plateau) where the dominant landscape encompasses desert sands interrupted with sandstone outcrops (Fig. 3.2). In addition, arable lands sometimes occur along the alignment between Bir Hermaz and Tabuk.

**Fig. 3.2** Desert sands and sandstone outcrops in the plateau of NEOM Region

### *3.1.4 Valleys*

NEOM Region, with its major geomorphologic units (coastal zone, mountain ridges and the plateau), is detached by several steephead valleys that span from the plateau along the mountains and then outlet in the coastal zone. Most of these valleys are found with gentle slopes (<5°) in the upstream area, and certainly where they are originated from in the plateau and then they become with steep slopes (25–30°) when they pass through the mountain ridges.

Except, Wadi Al-Roukieb and some other moderate and small valleys which spans eastward, most of the large valleys are trending initially westward and then to the south or south-west. These valleys are almost geologically controlled and they extend along different lithological units where, in many instances, the structure plays a role in their direction and orientation.

According to the available topographic maps (1:25.000 scale) obtained by the MoPMR (Ministry of Petroleum and Mineral Resources 1970), there are about 12 large valleys in NEOM Region where they are connected with 18 moderate valleys and there are also some doubled number of small ones. They are sometimes have different names as a results of the date when they were mapped and then the modified naming. These valleys can be classified as follows:

1. Large valleys: These are valleys with considerable dimensions where they extend for tens of kilometers and sometime they exceed 100 km such as the largest five valleys named (from north to south) as: Wadi Zieta, Wadi A'afal, Wadi Al-Roukieb, Wadi Sharma and Wadi Al-Kahlah.
   Thus, large valleys are characterized by wide width that average at about 250 m, while they have depth usually less than 5 m. The cross-section of these valleys are usually found as U-shape section, and thus indicating an old stage of valley formation. Whereas the flood plains of large valleys, which are usually wide (>100 m), are dominant with alluvial deposits mixed with soil and muds.
2. Moderate valleys: These valleys are usually connected with the large valleys at obtuse angles. They are often with less than 50 km length.
   Moderate valleys have width (50–250 m) and depth ranges between 3-8 m. Hence, the cross-section of these valleys are approximately with U-shape section where mature stage of development is evidenced. The flood plains of the moderate valleys often ranges between 50 and 100 m where alluviums are dominant.
3. Small valleys: These valleys usually compose reaches with length of less than 5 km (i.e. not usually observed on the available topographic maps). They are characterized by narrow cross-section (<50 m) and steep slopes with very wide (or negligible) flood plains; therefore, they sometimes found as gorges.

## 3.2 Climate

The region of the upper northwest of Saudi Arabia has a continental desert climate modified locally by proximity to the sea, and thus it receives localized rainfall mainly during the winter (Clark 1987). However, there is no doubt that the location of Neom Region is exclusive if compared with the rest parts of the Kingdom of Saudi Arabia. Thus, it is characterized by relatively cooler climate which is estimated at 10 °C temperature below the typical temperature across the Arabian Peninsula. This should be normal since NEOM Region, with its mountain ridges, builds a climatic barrier that capture the blowing wind and the associated wet clouds derived from the Red Sea.

Climate, with its two main variables the rainfall and temperature, is a significant physical parameters since it totally and sometimes partially controls many natural and anthropogenic components on terrain surface. Recently, climate has given attention and the issue of climate impact raised in context of the c climate change. However, the lack in climatic records still constitutes a constraints while applying hydro-climatic scenarios and projections.

According to the UNESCO classification (1979), the Arabian Peninsula is considered as 99% arid and hyper-arid zone, where rainfall rate is the lowest on the Earth. Thus, more than 90% of the Arabian Peninsula is characterized by annual mean temperature of about 20 °C, and some regions have annual mean temperature of more than 30 °C (Kotwicki and Al-Sulaimani 2009).

There is no climatic data dedicated for NEOM Region, and even the climatic data for the entire Tabuk Province is mainly dependent on the climatic station located in Tabuk City at latitude: 28° 22′ 21″ N; longitude: 36° 36′ 46″. This station is faraway at 50 km from the border of NEOM Region and 210 from the most remote locality in Neom Region at Ras Al- Shyiekh Hamid. Thus, the data available at this station is not representative to the entire NEOM Region except for the plateau area.

Therefore, data on climate for NEOM Region in this section has been adapted from different sources including mainly the following:

- World Weather (2020). Global broadcast and media distribution solutions.
- Tabuk Climatic Data. The General Authority of Meteorology and Environmental Protection (GAMEP 2019).
- Meteoblue. Climate Tabuk (Meteoblue 2020).
- Tropical Rainfall Mapping Mission (TRMM 2014).
- Climate Hazards group Infrared Precipitation with Stations (CHIRPS, 2015).

TRMM is a radar data (microwave) system executed by NASA and JAXA (Japan Aerospace Exploration Agency). It is able to illustrate rainfall measurements since 1998 up to 2015. This system retrieves rainfall datasets on daily basis and extend this data either as numeric records, table or even as maps. While, CHIRPS dataset built on smart interpolation techniques and high resolution, long period of record precipitation estimates based on infrared Cold Cloud Duration (CCD) observations.

### 3.2.1 Rainfall

NEOM Region is located above the latitude 27° 43′ 00″ N which distinguished it towards more precipitation rate if compared with the largest part of the Saudi Arabia, which does not receive much rain and the average yearly precipitation is approximately 100 mm/year.

In addition to the fossil groundwater, rainfall is considered as the main sources of water in the Kingdom of Saudi Arabia. Thus, rainfall distribution is not uniform in space and time and it is affected by several factor including the topography. Therefore, the Arabian Peninsula receives considerable amounts of rainfall in the northern, central-northern, and southwestern regions, whereas the southeastern region is almost entirely dry (Almazroui 2012).

Rainfall, as the major aspect of precipitation, frequently occurs between November and February reflecting the Mediterranean cyclones, while the rest months show negligible rainfall (Fig. 3.3). Besides, the rates of evaporation exceed many times the rates of precipitation since the area is characterized by arid climate (Al-Harbi 2010).

The average rainfall in Tabuk city (778 m) and the surrounding is about 47 mm/year as recorded from ground station in Tabuk City. While, the maximum and minimum reported rainfall rates between 1979 and 2008 was 80.6 mm and 1.1 mm/year; respectively (Hag El-Safi and El-Tayeb 2016). However, for the interior part of NEOM Region, the average rainfall rate was measured for Al-Sharaf Town (738 m) by using the TRMM and CHIRPS systems, and thus the average rate was about 57.5 mm/year (Fig. 3.3).

Using TRMM and CHIRPS system enabled determining the average rainfall rate in the crests zone (i.e. above 1600 m) which was found about 96 mm, which is almost

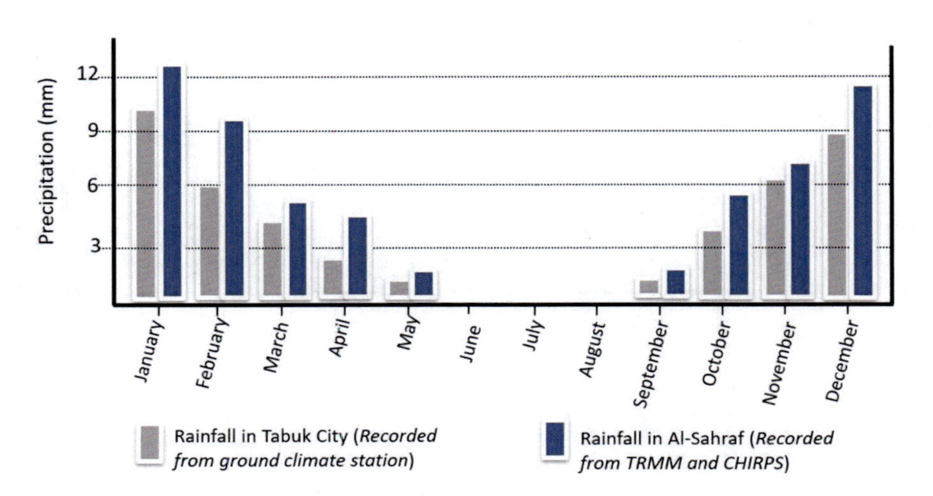


**Fig. 3.3** Average annual precipitation rates in Tabuk and NEOM Region

double the rate of rainfall in Tabuk City and reflects the influence of altitude factor. While the average rainfall rate was determined at about 51 mm in the coastal zone of NEOM Region. However, there negligible change in the rainfall rate when moving from south to north.

### 3.2.2 *Temperature*

Temperature remains the most important climatic parameter that characterize the entire Arabian Peninsula where arid climate is dominant with high temperature rates. The climate of the Kingdom of Saudi Arabia constitutes extreme aridity and heat. It is one of the regions in the world where temperature during summer reaches above 50 °C (Krishna 2014).

For Tabuk City and the surrounding, the estimated mean monthly temperature is about 23.3 °C (Fig. 3.4). While, the calculated mean monthly maximum temperature is 28.3 and 15.4 °C for minimum one. In addition, the highest temperature is between May and September (Fig. 3.4).

However, these measurements cannot be applied to NEOM Region which is characterized by different topographic features than Tabuk City. Therefore, mountain ridges in NEOM Region are supposed to have different temperature. In this respect, the temperature-altitude relationship was calculated depending on the fact that the temperature as a function of elevation is normally about 0.6 °C per 100 m. Therefore, three representative mountainous localities altitudes from NEOM Region were accounted by comparing their altitude with the altitude of Tabuk City. These are in Al-Sharaf (738 m), Jabal Al-Thaghra (1054 m) and Jabal Zuhed (1980 m). Thus,

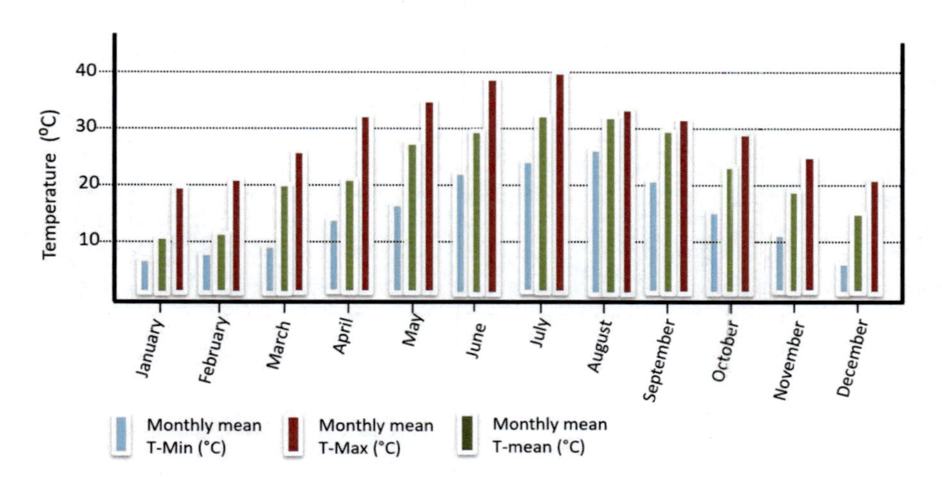


**Fig. 3.4** Mean monthly temperature and the maximum and minimum monthly temperature in Tabuk City

the temperature can be calculated as: 22.54, 19.57, 16.23 °C for Al-Sharaf, Jabal Al-Thaghra and Jabal Zuhed; respectively. Hence the average temperature in the mountain ridges of NEOM Region will be 19.05 °C. Besides, the temperature in coastal zone of NEOM Region has little difference with the temperature of Tabuk City. Thus, the monthly mean temperature, according to World Weather (2020), at Sharma City (10 m) is 24.30 °C and 24.65 °C in Al Bada'a (240 m). Therefore, the estimated average annual temperature in NEOM Region is about 22.5 °C, where the maximum reaches 44 °C and minimum 9 °C.

### 3.2.3 Climate Change

The territory of Saudi Arabia, including all natural and man-made components are vulnerable for climate change impacts, notably that the existing ecosystems are sensitive and easily changeable. Thus, the change climate is represented mainly by the increasing temperature which may raise the levels of evapotranspiration by about 1–4.5% at 1 °C increase and about 6–19.5% at 5 °C. In addition, yield losses of different types of field crops and fruit trees is anticipated between 5 and more than 25% (UNDP 2011).

The general understanding that there will be an increase in temperature and decrease in precipitation are expected to influence the agriculture and water supplies in Saudi Arabia. Nevertheless, there is still contradictory in the assessment of climate change magnitude. In this respect, scenario obtained by Al Zawad and Aksakal (2008) revealed that there is tendency for warmer temperature, more precipitation less evaporation, and more runoff, and hence this is allocated as: between 3 and 4.2 °C for temperature, 37.1 mm/year for precipitation, 20.8 mm/year for evaporation and 1.1 mm/year for runoff.

According to Meehl et al. (2007), t is anticipated to witness temperature increase between 2.5 and 5.1 °C, as well as the precipitation will increase by 30–41% in all regions of Saudi Arabia except the Northern region under. Other studies expected a decrease between 20 and 25% in precipitation over Saudi Arabia during the 2050s (Rajab and Prudhomme 2002).

For NEOM Region, the assessment of changing climate is viewed on wide geographic scheme and not only for the region limits. Therefore, the following evidences can be summarized:

- For Tabuk Province, the trends of annual, seasonal and even monthly temperature (i.e. mean, maximum and minimum) time series were investigated over 30 years between 1984 and 2013. The results anticipated significant rise in temperature by about 1.93 °C (Krishna 2014).
- According to Hag El-Safi and El-Tayeb (2016), the statistical rainfall analysis applied to 20 climatic stations in Saudi Arabia between 1979 and 2008 shows that the average rainfall rate in Tabuk almost remained stable between 1979 and 991,

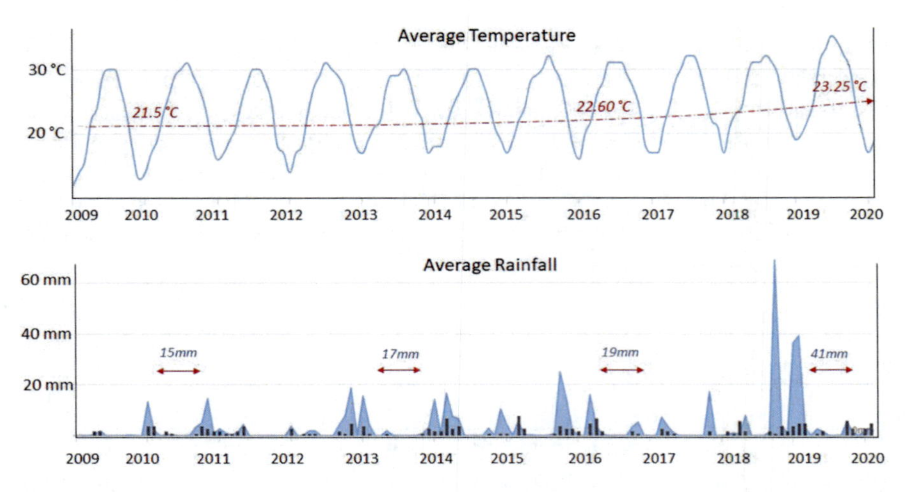


**Fig. 3.5** Example of the changing temperature and rainfall in Al Bada'a between 2009 and 2020 (World Weather 2020)

and then it has been decreases to 50 mm, 27 mm and 18 mm in 1994, 1997 and 2006; respectively.

According to World Weather (2020), there is an increase in the average temperature between 2009 and 2020 from 21.5 to 23.25 °C which was recorded for Al Bada'a (Fig. 3.5). While, the average rainfall, for the same region, is also increased from about 15 mm in 2009 to about 41 mm in 2020 (Fig. 3.5).

## 3.3 Hydrology

As a major component of the hydrological assessment, the flow regime of the surface water resources will be discussed in the next chapters of this book where maps are generated. This will include the discussion and illustrations for drainage system analyses (i.e. watersheds and stream networks and their geometry and morphometry) will be elaborated; however, this section will focus on the existing surface water resources in NEOM Region in spite of that there is very little data on the hydrology of this region. Therefore, building an inventory on the existing surface water resources would be supplementary for the identification of the physical characteristics of this region which supposed to become a global zone.

NEOM Region can be considered as an area with little surface water resources due to its climate which is almost arid and the rainfall rate does not exceed 60 mm, with excessive evaporation rate.

### 3.3.1 Springs

Even though springs are often considered as groundwater resources, yet the author allocates them with surface water due to the fact that springs are observable water resources that combine between water flow/storage regime whether on terrain surface or in subsurface.

In Saudi Arabia, there are several springs known since the ancient times where the inhabitants used them for domestic irrigation purposes. Lately, most of the springs located in the central and eastern areas of Saudi Arabia become dry. However, no studies done on springs located in the northwest region of the country.

No major springs can be found in NEOM Region, but there are some major ones located outside the region, like springs of Ain Sukker in Tabuk, and Ain Ed-Dissah (45 km southwest of NEOM Region).

Based on the field surveys and the topographic maps (1:250.000 scale) produced by the Ministry of Petroleum and Mineral Re-sources (MoPMR 1970), there are few number of seeps and springs found in NEOM Region.

The major recognized springs are:

- Al-Nejieleh (Lat. 28° 55′ and Long. 35° 36′; 1450 m; sand-stone rocks).
- Al-Katar (Lat. 28° 54′ and Long. 35° 30′; 1280 m; sandstone rocks).
- Spring near Al-Sharaf (Lat. 28° 55′ and Long. 35° 10′; 713 m; alluvial deposits).
- Spring near Aafal (Lat. 28° 43′ & Long. 35° 06′; 502 m; alluvial and colluvial deposits).
- Fehyman (Lat. 28° 05′ and Long. 35° 26′; 255 m; diorite).
- Ain Qammah (Lat. 28° 34′ and Long. 35° 18′; 1490 m; clastic rocks and slate and phyllite).
- Ain Nimah (Lat. 28° 07′ and Long. 35° 42′; 1220 m; granite).
- Bir Al-Zereb (Lat. 28° 05′ and 35° 33′; 1450 m; alluvial and colluvial deposits).

Most of these springs are characterized by low discharge (less than 10 l/s), and recently most of them have been dried. Therefore, inhabitants follow harvesting approaches in order to conserve water seeps with low flow from these spring. Therefore, the inhabitant construct rounded concrete ponds (few meters in diameter) to collect water from these springs (Fig. 3.6). This is well pronounced in the coastal zone where the located springs there are issuing from the alluvial and colluvial deposits. These pond are open at surface in order to be used for livestock.

As a common hydrologic phenomenon, the geothermal water is anticipated in NEM Region. The hydrologic clues behind this concept is that the region occupies several plutonic rocks which serve in heating groundwater. This needs to be investigated in-depth, notably there are some geothermal springs located along the western Saudi coast like Al-Lieth Spring (150 km south of Jeddah), and a geothermal spring in Jizan.

**Fig. 3.6** Spring issuing in concrete pond near Al-Sharaf town

### *3.3.2 Wetlands*

The localities where water accumulates for considerable time periods and retain the soil saturated are described as wetlands. These wetlands are usually characterized by specific ecological systems and become habitats for many plant and animal species. They are well known in many regions of Saudi Arabia where they exist as Sebkhas, mudflats, mangroves, and valleys (Al-Obaid et al. 2016).

Even though NEOM Region is known by few surface water sources and very low precipitation rate (<60 mm), yet surface water bodies can be found there. These bodies are temporally exist either as:

- Terrestrial wetlands where they are found along valleys courses and particularly where flood plain occur with low-lands and depressions. In addition, some surface water bodies are originated between mountain hills. This type of wetland usually has a dimension ranges between 0.2–0.4 km$^2$.
- Marine wetlands which are dominant in the coastal zone where they occur as sebkhas and mangroves.

The terrestrial wetlands retain water for few months (2–4 months) after rainfall takes place, while the soil there remains saturated for similar time periods. Therefore, locals living nearby these localities, where rainfall water accumulates for few months, try to benefit from this water; therefore, they often construct retaining walls or small dam to harvest as much as they can from this water, which is used mainly for irrigation purposes and sometimes for livestock.

The localities where these surface water bodies (i.e. wetlands) are sometimes identified and either they are named by locals. These wetlands can be also recognized from satellite images which are retrieved at least one month after raining periods.

The most known surface water bodies (i.e. terrestrial wetlands) in NEOM Region (or almost in its surrounding) are plotted on the topographic maps (1:250.000 scale) where they are named as "Ka'a" and "Mshash". These are (with their approximate coordinates and altitude):

- Mshash Al-Ras (35°19′ and 28° 50′; 1058 m)
- Mshash Khodiery (35° 03′ and 28° 58′; 1054 m)
- Mshash Zuryek (34° 58 and 28° 41′; 673 m)
- Ka'a Amem (35° 41′ and 28° 40′;1410 m)
- Ka'a Um Sali (35° 43′ and 28° 31′; 1252 m)
- Ka'a El-Mohrek (35° 50′ and 28° 43′,1095 m)
- Mshash Dum (35° 48 and 28° 28′, 1285 m)
- Mshash Bou Dress (35° 48′ and 28° 18′; 975 m)
- Ka'a Al-Kammam (35° 55′ and 28° 08′; 965 m)
- Mshash Raymeh (35° 25′ and 28° 13′; 1130 m).

With respect to the marine wetlands, there are two major sebkhas. These are Al-Seeh (near Al-Khreybeh bay), with 3 km × 0.5 km; and Al-Akooz (near Ayeynat), with 2 km × 4 km; whereas, the mangrove is almost spread over about 65% the southern coast of NEOM Region.

### *3.3.3 Snow*

It is a paradox to include solid precipitation (i.e. snow) in the hydrological studies of the Arabian Peninsula, the arid zone with large geographic territory covered by desert. In particular, NEOM Region is witnessing this phenomenon, while large area of NEOM is a desert and the rainfall rate does not exceed 60 mm. However, snowpack is accumulated on several crests in NEOM Region and this might be every year, and this added considerable amounts to the water budget.

As a newly hydrologic phenomenon, there are many other aspects of solid precipitation have been also occurred in NEOM Region. This includes hail, sleet, graupel, etc. and many forms of ice crystals and pellets. Nevertheless, these solid and frozen water form almost covers large area if compared with snowpack; and therefore, they spread over different regions whether at high or low altitudes.

In contrast, snowpack is distributed on the crest of NEOM Region and this has been witnessed many time over the past decades, but it has been well noticed lately as a function of the changing climate in the region as a whole. Thus, snowpack in NEOM Region is almost observed on the altitudes above 1800 m where it is typically covering Jabal Al-Louz (2401 m), Jabal Al-Kloom (2398 m) and Jabal Zuhed (1980 m).

Remarkably, snow covered large area of NEOM Region in January 2020, as observed in Modis Terra satellite images. Hence, snow covered about 2850 km2 of NEOM Region, which is equivalent to about 11% of NOEM. Therefore, snowpack of January 2020 almost covered all crests above 1550 m, and thus, it raised up 2100 m after one week as it was observed on satellite images.

**Table 3.1** Dams in NEOM Region

| Dame name | Coordinates | | Dimensions | | Capacity (million $m^3$) |
|---|---|---|---|---|---|
| | Latitude | Longitude | Length (m) | Height (m) | |
| Wadi A'afal | 28° 56′ 29″ | 35° 10′ 49″ | 401 | 12 | 1.52 |
| Wadi Sharma | 28° 05′ 03″ | 35° 30′ 07″ | 1112 | 10 | 1.0 |
| Wadi Aynona | 28° 12′ 39″ | 35° 19′ 41″ | 400 | 9 | 0.71 |
| Al-Muieleh | 27° 35′ 16″ | 35° 32′ 23″ | 168 | 12.5 | 1.54 |
| Wadi Dumm | 28° 34′ 07″ | 35° 59′ 59″ | 510 | 10 | 0.85 |

### *3.3.4 Reservoirs*

Reservoirs are artificial lakes that are created when dams of different dimensions are constructed. They are executed for several functions including mainly surface water harvesting, controlling floods, artificial groundwater recharge, fisheries and establishing promenades sites with distinguished landscape. Therefore, when talking about reservoirs, it is usually described from the concept of dams building.

Dams in the Kingdom of Saudi Arabia are tremendous and they have been adopted recently as measures for flood controls, but they are also utilized for collecting rainfall water that flows along wadis. Hence, Saudi Arabia constructed hundreds of dams in the last 10 years where they are characterized by different dimensions, purposes, and engineering specifications.

Up to 2020, there are 11 dams constructed in Tabuk Province, and there is almost similar number of proposed ones. The total capacity in the reservoirs of the constructed dams is about 9.5 million $m^3$. However, in NOEM Region, as a part of Tabuk Province there are five main dams constructed to retain the flowing water in wadis and thus capture considerable amount of water in the adjacent reservoirs to control floods as a primary function. Hence, these reservoirs have a capacity exceeding 5.5 million $m^3$. These dams and their specifications are shown in Table 3.1.

### *3.3.5 Ground Ponds*

These are man-made constructions obtained to store water which can be supplied either from springs or from drilled boreholes which are dug in shallow groundwater reservoirs. It is common features applied in NEOM Region, and more specifically in the coastal zone where they are wide spread.

These ponds are rounded in shape, closed on surface and they are usually with approximately 5 m diameter and 3 m height and the estimated capacity exceed 50 $m^3$. They are constructed as concrete ponds where they are found on terrain surface. These ground ponds are executed either on hills or on soil dumps in order to convey water by gravity, or on they can be on flat lands but water pumps are fixed then to abstract water from the ponds and convey it for different purposes including mainly domestic ones.

These ponds are often found as clusters (i.e. several ponds in small area) in the coastal towns where some of them are abandoned due to water leakage or the depletion of groundwater sources.

They have been known along the southern coast of Tabuk Province (recently NOEM Region) since 1960s, and the most exploited ponds were even plotted in the topographic maps produced by the MoPWR (1970). Therefore, the most popular ponds are: Abar Abo-Za'ala; Abar Ghanem, Abar Al-Osieli, Abar Qayal. Whereas, the term "Abar" in Arabic means boreholes, and they named so because these ponds are filled with water from the boreholes which are dug in the shallow groundwater reservoirs contained in the alluvial deposits along the coastal zone of NEOM Region.

## 3.4 Geology

The Kingdom of Saudi Arabia can be divided into two main geological provinces. These are: the western province which is composed largely of crystalline igneous and metamorphic rocks that are tilting to the northeast, east and southeast and belong to Precambrian age. While, the eastern province constitutes sedimentary rocks overlying the basal igneous and metamorphic complex and these rock are in general inclined to the east.

There are a number of studies applied on the geology of the Western Arabian Peninsula. In this respect, field and laboratory investigations for exploration purposes have been initially carried out by USGS-ARAMCO (1963). Therefore, the MoPMR in the Kingdom of Saudi Arabia has performed geological mapping of the country and divided the territory of the Kingdom into quadrangles (i.e. geologic map sheets) with define area and coordinates. For each quadrangle, the mapping including field surveys and reporting, has been carried out by several geologists, such as Brown et al. (1989) and Moore and Al-Rehaili (1989).

The Western part of the Arabian Peninsula constitutes the major rock body of the Precambrian shield, which is named as the Arabian Shield. This shield is bounded from the west by the Red Sea Rift System where the Paleozoic and the overlying sedimentary rocks extend in the east. Therefore, NEOM Region is located in the most upper part of the Precambrian Arabian Shield.

In order to cover the entire NEOM Region, there are three quadrangles used with scale of 1:250.000. These are:

- Haqel map (sheet: 29 A), by Rowaihy (1985)
- Al- Bada'a map (sheet: 28 A), by Clark (1987)
- Al-Muieleh map (sheet: 27 A), by Davies and Grainger (1985).

During the analysis of the three geologic maps as well as several related studies belong to the geology of the NEOM Region, contradictory has been found in the nomenclatures and classification of the exposed rock formations. However, the three obtained geologic maps their explanatory reports obtained by Rowaihy (1985), Clark (1987) and by Davies and Grainger (1985) were adopted.

### 3.4.1 Stratigraphic Sequence

Based on the geological maps (1:250.000) mentioned previously, the stratigraphic sequence for NEOM Region was studied in-depth and this was accompanied with field investigations. In this respect, Al-Bada'a map (sheet: 28 A) was the basic document used. A simplified illustration of this map is shown in Fig. 3.7.

The majority of the stratigraphic succession of NEOM Region constitutes Cenozoic, Mesozoic and Paleozoic and Proterozoic eras. These eras compose of 80 rock formations starting from Precambrian to Recent. These rock formations and their lithologies can be summarized (from oldest to youngest) as follows:

1. Proterozoic

   - Overlapping rocks and complexes:

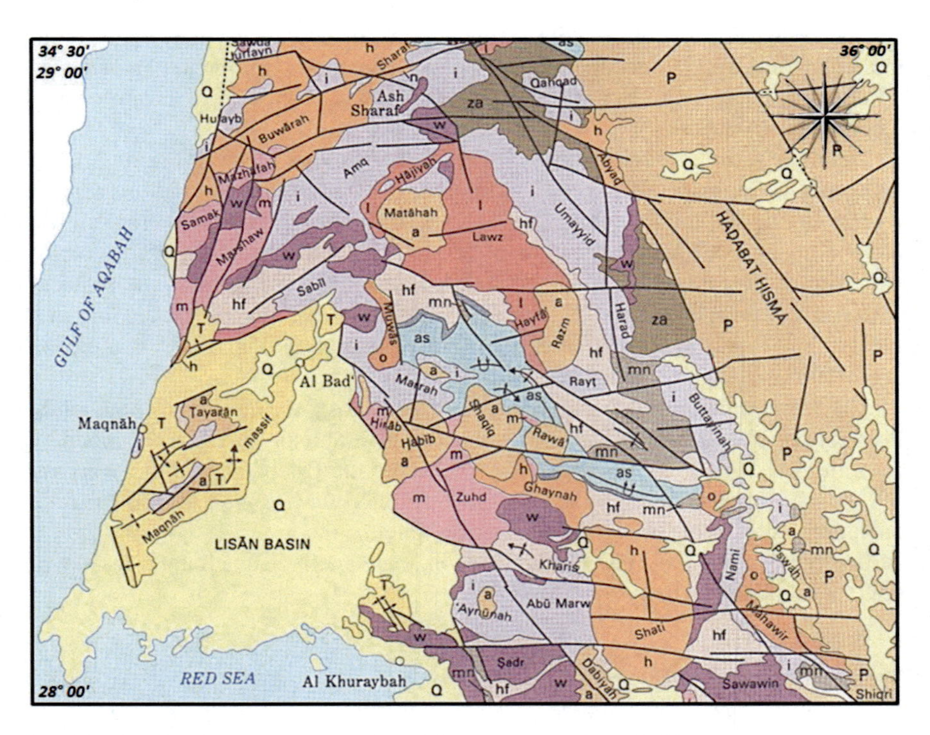


Fig. 3.7 Simplified geologic map of Al-Bada'a quadrangle (Clark 1987)

- Hegaf Formation (hf): Metamorphic volcanics and subordinate sedimentary rocks, mafic lithologies, with local iron ores and marble.
- Silasia Formation (si, zs, zsd): Metamorphosed greywacke, siltstone, andesite tuff, jaspilitic; Tuffaceous sedimentary rocks; iron ores and diabase sills.
- Zaytah Formation (za): Metamorphosed silicic lava and tuff, greywacke, metamorphosed felsic schist, amphibolite and marble.
- Hinshan Foramtion (hn, hms, hpy, hr, hrd, hva): Rhyolit, andesite and volcaniclastic rocks; greywacke and siltstone; andesite agglomerates; rhyolite; rhyodacite; andesite and tuff.
- Amlas Formation (as): Clastic sedimentary rocks, slate and phyllite, subordinate andesite, silicic tuff.
- Minaweh Formation (mn): Silicic lava and tuff, breccia, ignimbrite, conglomerates, sandstone and shale.
- Muklar Complex (kum): Ultramafic rocks, highly altered.
- Muwaylih Suite (wndi, wsdt, wst): Sawsan complex: diorite, quartz and gabbro; Sader complex: diorite, gabbro; trondhjemite.
- Ifal Suite (iqmg, iag, ijgd, idg): Monzogranite and granodiorite; monzogranite, biotite Jurfayn complex: granodiorite, monzogranite and quartz, momzodiorite and hornblende.
- Mabrak Granite (mmg): Biotitie, monzogranite with subordinate syenogranite.
- Ghadiyah Granite (gg): Hornblende syenogranite with subordinate quartz syenite.
- Asmar Comples (ato, aqd): Tonalie; quartz diorite.
- Quartz Monzonie (qm): quartz and monzonite.
- Atiyah Monzogranite (amg): Monzogranite, equigranular to porphyritic.
- Mowasse Quartz Syenite (oqs): Alkali-feldspar-quartz syenite to quartz syenite and Alkali-feldspar syenite.
- Midyan Suite (mzag, mdag): Zuhd alkali granite: alkali granite, minor alkali quartz syenite and alkali microgranite; Dabbagh comlex: alkali granite.
- Haql Suite (hgr, hbfg, lbg, htsg, hgfg, lrg, lag,): Granophyre and alkali-fledspar granite; Buwarah granite: alkali-feldspar granite to syenogranite; granite; Shati granite: syenogranite to alkali-feldspar granite; Ghanah comples: alkali-feldspar granite and syenogranite hornblende bearing; Rama granite, alkali-feldspar granite; Ayyad granite.
- Wasit Granite (wg): monzogranite with subordinate syenogranite.
- Lawaz Complex (lsg, lmg): Syenogranite, monzogranite and granophyre; Monzogranite to s.
- Unassigned intrusions of various ages (gr, qs): Monzogranite, syenogranite and microgranite; quartz syenite to alkali feldspar granite.
- Sawda Complex (nsy, sas): alkali-feldspar syenite and nepheline syenite; nepheline syenite and alkali-feldspar syenite.
- Duba Complex (dg, dt): Granodiorite, hornblende-bearing; tonalite.

  - Shar Complex (tsg): Alkali granophyre.
  - Maharishi Complex (mmg, mmt): Granodiorite and Tanolite porphyritic; intrpultonic, synplutonic syenite.

- Unassigned Plutons:
  - Alkali-granite (agr).
  - Syenogranite (gs).
  - Monzogranite (gm).
  - Granodiorite (gd).
  - Gabbronorite (gn).
  - Diorite (di).
  - Granbbomonrite to diorite (Tg) dikes.

2. Paleozoic
   - Cambrian:
     - Siq Sandstone (Cs): Local basal conglomerates, dark red.
     - Quwiera Sandstone (Cq): Local basal conglomerates, subordinate quartz-pebble conglomerate and silty sandstone.
   - Cambrian-Ordovician:
     - Ram and Umm Sahm Sandstone (OCr): Subordinate quartz-pebble conglomerate and sandy shale.
   - Ordovician-Silurian- Devonian:
     - Tabuk Formation (DOt): Sandstone, siltstone and shale, minor limestone.
   - Cretaceous—Tertiary (Eocene):
     - Adaffa Formation (TKa): Sandstone, marl, and shale, subordinate fossiliferous chert and limestone, minor conglomerate.

3. Cenozoic
   - Oligocene:
     - Sharik Formation (Ts): Conglomerate and sandstone, subordinate siltstone.
     - Azlam Formation (TKa): Basal sandstone, fossiliferous, argillaceous limestone.
   - Miocene (Raghama Group):
     - Raghama Formation (Tr): Basal reef limestone, calcareous siltstone and sandstone.
     - Musayr Formation (Trm): Sandstone and conglomerate, reef and sub-reef limestone and gypsum.
     - Nutaysh Formation (Trn, Trnm): Conglomerate, sandstone, marl and limestone; marl and subordinate gypsum, sandstone and siltstone; Trnm: Marl and subordinate gypsum, sandstone nad siltstone.

  - Bad'a Formation (Trb): Gypsum and anhydrite, subordinate marl, siltstone, sandstone and limestone.
- Pliocene-Pleistocene:
  - Lisan Formation (QTi): Sandstone and conglomerate, sand and local gypsum, poorly consolidated.
  - Conglomerate (QTc): Regolith of ancient terrace; gravel sheets and residual pebbles and boulders.
- Pleistocene-Holocene:
  - Gravel sheets and terraces (Qg).
  - Raised reef limestone (Qr).
  - Reef and sub-reef limestone and saline sand (Qml).
  - Sabkha (Qsb).
  - Aeolian sand dunes and sheets (Qes).
  - Undifferentiated sand and gravel and scree deposits (Qu).
  - Wadi alluvium (Qal).
  - Terrace deposits (Qt).

### *3.4.2 Rock Structures*

The upper northeastern part of the Arabian Peninsula is dominant with rock deformations which are influenced by different (i.e. dimension, type and date) tectonic forces, and then resulted complicated geologic structures, in particular the faults and folds.

In accordance with Clark (1987), the majority of lithological distribution of Al-Bada'a quadrangle, where NEOM Region is situated, can be divided into three principal lithological blocks as shown in Fig. 3.8. These are:

- Lisan Basin: It is located in the most south-western part of the region and forms delta-liked shape with low-lands where Quaternary deposits are dominant.
- Massive mountains: This zone, which is morphologically compose the mountain ridges, occupies the complex geology of NEOM Region where the age of rock formations is between Proterozoic to Quaternary. It constitutes (by number) more than 90% of the rock formations in the area.
- Hadabat Hisma: It is the plateau area that located in the north-eastern of NEOM Region, and it is composed from the Paleozoic sandstone rock formations that spread between sand dunes.

Faults:

There are several fault systems occur in the area of Al-Bada'a quadrangle, where NEOM Region is located. These faults are found as deformation sets with define directions indicating the impact regional tectonism. Thus, the entire region lies in the

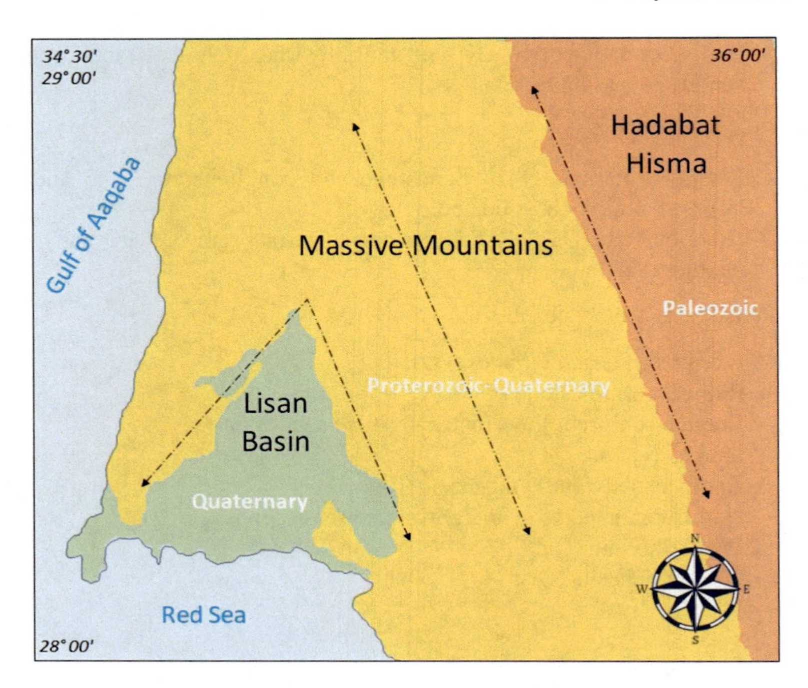


**Fig. 3.8** Schematic map for the major lithological blocks in NEOM Region

proximity of the intersection between three main fault systems (Clark 1987). These are: (1) the Najd system, trending northwest, (2) the Red Sea Rift System, trending north- northwest and (3) the Gulf of Aaqba system, trending north to northeast. In addition many east-trending faults, but with less extent, are present. Most of these faults are probably date from the Late Proterozoic and were reactivated in the Tertiary.

The area of study is just north of the major Late Proterozoic Najd wrench-fault system (Moore 1979) along which a cumulative left-lateral movement of about 240 km has been occurred (Brown 1972). Thus, Al-Muieleh fault zone, as a significant structural boundary, extends northerly to Najd structure. There is also the west-northerly striking fault south of Jabal Zuhed.

Faults of the Tertiary Red Sea System are present throughout most of the region, i.e. fault zone at Jabal Zuhed and Jabal Al-Shiqri which are associated with dike swarmers. These are Gabbroic dikes of probably Tertiary age (Blank 1977) which are present along fault traces in several places (Fig. 3.9).

Faults belong to the Gulf of Aaqba are most clear within a few kilometers of the coasts. While, the total movement along the Gulf of Aaqba-Dead Sea Rift System is estimated to be about 105 km (Dubertret 1932).

Fault drag along the Gulf of Aaqba fault zone can be observed by the conspicuous changing in direction of dike swarmers from essentially E-W away from the Gulf to N-S close to it.

**Fig. 3.9** Composite and mafic dikes intruded in Ifal Suite (i.e. monzogranite and granodiorite)

Other faults of variable orientation are also present in the region and probably date from the Proterozoic. Minor thrust faults are associated with the Muklar Complex (i.e. ultramafic rocks), but no major faults of this type have been observed (Clark 1987).

1. Folds

The intrusions of large granite plutons resulted Proterozoic stratiform rocks which are highly disjointed outcrops patterns. Nevertheless, an overall westerly to north-westerly structural grain is evident. However, the existed three volcano-sedimentary sequences are probably associated with separate tectonic events (Clark 1987).

There are a number of folding systems occur in the area of study (Clark 1987). These are found in:

- Hegaf and Silasia Formations, which reveal sub-vertical foliation consisting of parallel bedding, implying isolinal folding.
- Amlas Formation composes a synclinal structure that overturned along the southern contact with the Hegaf Formation.
- Minaweh Formation where beds dip gently and sometimes moderately and are openly folded.

The Proterozoic rock formations in the eastern part of the Al-Bada'a quadrangle are generally tilted and sometimes faulted but are otherwise they are unreformed. In addition, block faulting largely controlling the sedimentation and deformation of the Mesozoic and Cenozoic rock formations (Clark 1987).

## References

Al Zawad, F., & Aksakal, A. (2008). Impacts of climate change on water resources in Saudi Arabia. In *The 3rd International Conference on Water Resources and Arid Environments and the 1st Arab Water Forum*.

Al-Harbi, K. (2010). Monitoring of agricultural area trend in Tabuk region—Saudi Arabia using Landsat TM and SPOT data. *The Egyptian Journal of Remote Sensing and Space Science, 13*(1), 37–42

Almazroui, M. (2012). Temperature variability over Saudi Arabia and its association with global climate indices. *Meteorology, Environment and Arid Land Agriculture, 23*(1), 85–108.

Al-Obaid, S., Samraoui, B., Thomas, J., & O'Connell, M. (2016). An overview of wetlands of Saudi Arabia: Values, threats and perspectives. *Ambio, 2017*(46), 98–108. https://doi.org/10.1007/s13280-016-0807-4.

Blank, H. (1977). Aeromagnetic and geologic study of tertiary dikes and related structures on the Arabian margin of the Red Sea. Saudi Arabian General Directory of Mineral Resources. Bulletin 22.

Brown, G. (1972). Tectonic Map of the Arabian Peninsula. Saudi Arabian Directorate General of Mineral Resources, Arabian Peninsula Map, AP-2, scale 1:4.000.00.

Brown, G. F., Schmidt, D. L., & Huffan Jr, A. C. (1989). *Geology of the Arabian Peninsula, Shield area of Western Saudi Arabia*. U.S. Geological Survey Professional Paper, 560-A.

CHIRPS. (2015). *Climate hazards group infrared precipitation with station data*. Available at http://chg.geog.ucsb.edu/data/chirps/.

Clark, M. (1987). *Geologic map of Al-Bada'a quadrangle, A-28; (1:250.000)*. Ministry of Petroleum and Mineral Resources.

Davies, F., Grainger, D. (1985). *Geologic map of Al-Muieleh quadrangle, A-27; (1:250.000)*. Ministry of Petroleum and Mineral Resources.

Dubertret, L. (1932). *Les forms structurales de la Syrie et de la Palestine; leur origine*. (Vol. 195, pp. 65–67). Academie des Sciences Comptes Rendus.

GAMEP (The General Authority of Meteorology and Environmental Protection). (2019). *Tabuk climatic data*. Available at https://www.pme.gov.sa/Ar/Meteorology/Pages/ClimateReport.aspx.

Hag El-Safi, S., El-Tayeb, M. (2016). *Spatial and statistical analysis of rainfall in the Kingdom of Saudi Arabia from 1979 to 2008* (Vol. 71, No. 10, pp. 262–266). *Royal Meteorological Society*. Weather—October 2016. https://doi.org/10.1002/wea.2783.

Kotwicki, V., & Al-Sulaimani, Z. (2009). Climates of Arabian Peninsula, past, present and future. *International Journal of Climate Change Strategies and Management, 1*(3), 297–310.

Krishna, L. (2014). Long term temperature trends in four different climatic zones of Saudi Arabia. *International Journal of Applied Science and Technology*, *4*(5), 10.

Meehl, A., Stocker, T., Collins, W., Friedlingstein, P., Gaye, A., Gregory, J., et al. (2007). Global climate projections. In: *Climate change*. NY, USA: 4th Assessment Report of the Intergovernmental Panel on Climate Change.

Meteoblue. (2020). *Climate Tabuk*. Available at https://www.meteoblue.com/en/weather/historyclimate/climatemodelled/tabuk_saudi-arabia_101628.

Moor, A., & AL-Rehaili, M. (1989). *Geological map of Makka quadrangle sheet 21D*. Kingdom of Saudi Arabia: Ministry of Petroleum and Mineral Resources.

Moore, J. (1977). Tectonics of the Najd transcurrent fault systems, Saudi Arabia. *Journal of Geological Society of London, 136*, 441–454.

MoPWR (Ministry of Petroleum and Mineral Resources). (1970). *Topographic maps of Saudi Arabia, 1:250.000*. Riyadh, KSA: Aerial Survey Department.

Rajab, R., & Prudhomme, C. (2002). Climate change on water resources management in arid and semi-arid regions: prospective and challenges for the 21st century. *Biosystems Engineering, 81*(1), 3–34.

Rowailhy, M. (1985). *Geologic map of Haqel quadrangle, A-29; (1:250.000)*. Ministry of Petroleum and Mineral Resources.

TRMM (Tropical Rainfall Mapping Mission). (2014). *Rainfall archives.* NASA. http://disc2.nas com.nasa.gov/Giovanni/tovas/TRMM_V6.3B42.2.shtml.
UNDP. (2011). *Second national communication Kingdom of Saudi Arabia* (209 pp). Submitted to: The United Nations Framework Convention on Climate Change (UNFCCC)
UNESCO. (1979). Map of the world distribution of arid regions: Map at scale 1:25,000,000 with explanatory note. MAB Technical Notes 7. Paris: UNESCO.
USGS-ARAMCO. U. S. Geological Survey-Arabian American Oil Company. (1963). Geologic map of the Arabian Peninsula: U.S. Geological Survey Miscellaneous Geologic Investigations Map I-270 A, scale 1:2,000,000.
World Weather. (2020). Global broadcast and media distribution solutions. In *Al Bad' monthly climate averages.* Available at https://www.worldweatheronline.com/al-bad-weather-averages/tabuk/sa.aspx.

# Chapter 4
# Thematic Maps

**Abstract** Usually production of maps is a constraint occurs when projects are planned and intended to be achieved. The majority of this constraint implies the lack of data and expertise as well tools needed. Thus, maps are usually the simple and rich documents that enable stakeholders of different levels; e.g. individuals, experts, decision makers, etc. to read and understand these documents where comprehensive and thematic data are present. Therefore, maps are often considered as the basic data and information required for SLM projects. It is a typical example in this study where NOEM Region, as a bare land, is proposed to be a Global economic hub. No remarkable activities have been done in this region before, and it is still uncovered and no reliable data were found for this region, notably the thematic maps which are used as supplementary geo-spatial data. Therefore, planning approaches need to assure primarily the availability of these maps. This chapter will illustrate the techniques used for maps production applied for NEOM Region. They are the first of their type done for NEOM Region. The generation of these maps followed new advanced techniques of data acquisition, extraction of terrain elements and the manipulation of geo-spatial data. For this purpose, space techniques (certainly satellite images) and geo-information systems were utilized in an integrated mapping approaches. Thus, the produced maps include valuable information and measurements on different themes for NEOM Region. This chapter does not aim just to produce these maps, but also to illustrate the results of measurements for the investigated themes.

**Keywords** Topography · Drainage density · SPOT images · Temporal resolution · DEM · Arc-map

## 4.1 General Overview

Mankind has long recognized the significance and value of maps to his life. In fact, the history of mapping can be referred to more than few thousands of years ago. Until the last three decades, maps production was following traditional methods whether on the retrieve data and information on terrain surfaces and the existing components or in the ways by which these maps were drawn.



M. M. Al Saud, *Sustainable Land Management for NEOM Region*,
https://doi.org/10.1007/978-3-030-57631-8_4

Recently, mapping is represented principally by the topographic maps or charts. These maps are characterized by large-scale and variety of quantitative representation of vertical and horizontal dimensions of land surface which are described as contour lines (i.e. connection between points of equal elevation).

The generation of topographic maps was primarily obtained from aerial photographs (or airborne imagery) which were taken from an aircraft or any flying object and these photographs are characterized by several specifications and approaches of interpretation. Thus, the first recorded example of a photograph taken from an airplane by Wilbur Wright in 1909. This led to the development of photogrammetry which enables making measurements from photographs, drawing, a measurement, or digital elevation models.

A topographic map is the base document (or base map) upon which the thematic information exist on terrain surface can be plotted. Therefore, for any map generation, the topographic map is primarily used and then the themes of concern (e.g. geology, geomorphology, agriculture, etc.) are projected on its. Therefore, thematic maps are produced to perform the geo-spatial information of a specific theme.

Digital and photographic satellite image data offer a high potential of topographic and thematic information where the application of such data for mapping purposes. Therefore, satellite remote sensing has been used for topographic mapping purposes since the launch of ERTS-1 in 1972 (Dixon-Gough 1994).

The Kingdom of Saudi Arabia started performing surveying and mapping since the 1950s when the Ministry of Petroleum and the Ministry of Defense initiated their topographic mapping program at the scale 1:50,000 (Alrajhi 2005).

The available topographic maps in this study were obtained by the Aerial Survey Department, related to MoPMR in 1970. These base maps have the following specifications:

- Scale 1:250.000.
- Contour interval 50, and 25 m in flat area.
- Magnetic declination for 1965, 2° easterly for the entire area.
- Projection: projection universal transverse Mercator. International spheroid.
- National Geodetic Net (1970).
- Vertical datum: Mean sea level, Jeddah (1969).
- Produced from aerial photographs taken in 1970.
- Produced by Pacific Aero Survey Co., Ltd.

Recently, the use of space techniques has become widely spread in several regions and for different applications, including the topographic maps. In a broad sense, it tackle the processing of digital (i.e. electronically produced) satellite images, which are significant instruments used to draw maps, identify, analyze and measure the observable terrain features that are reflecting several natural and anthropogenic processes.

## 4.2 Remote Sensing Techniques

Remote sensing is the process of data acquisition by satellite (or aircraft). This data represents the reflected and emitted electro-magnetic radiation (EMR) from Earth's surface. This radiation (or energy) enables identifying the spectral signatures which are reflected as diverse objects with different physical characteristics.

Since the first lunched satellites (e.g. Sputnik, 1957; Explorer, 1958 and Corona, 1960) satellite remote sensing has been widely used and attempts for innovative methods became common. Thus, remote sensing has a very wide range of applications in many different fields, such as coastal zones management, agricultural, water resources, natural hazards, change detection, marine pollution, etc.

This electro-magnetic energy is considered to span the spectrum of wavelengths from 10–10 mm to cosmic rays up to 1010 mm, and the broadcast wavelengths, which extend from 0.30 to 15 mm.

There are two types of remote sensing instruments (i.e. sensors), and they are either passive (optical) or active (microwave). Passive sensors detect only natural radiation reflected or emitted by the object from a source. While active sensors provide their own energy source for illumination which is directed toward the target to be investigated, and then it detects the radiation reflected from that target.

Based on the type of radiation acquired by sensors and the wavelength regions; therefore, remote sensing can be categorized as:

- Visible and Reflective Infrared RS,
- Thermal Infrared RS,
- Microwave (or radar) RS.

Sensors are mounted on platforms (i.e. settles) which are characterized by different specifications and mainly the altitude above Earth's surface and the orbiting time. These platforms can be as: (1) satellites (500–900 km), (2) space shuttle (185–575 km), (3) high-latitude flying aircraft (10–12 km), 4) moderate-latitude flying aircraft (1.5–3.5 km) and 5) low-latitude flying aircraft (below 1 km).

Therefore, radiation from objects on Earth are acquired by sensors fixed on satellites, spacecraft, airplanes (with tremendous types), and lately drones. Hence, the flight characteristics of the platforms are significant for the quality of the retrieved geos-spatial data, but the most significant is the sensor characteristic themselves. However, satellites has recently given more attention because they can cover large areas with least cost. Thus, there are thousands of satellite orbiting around the Earth where considerable number of them are observatory satellites which are used to study the processes and properties of objects on Earth's surface.

### 4.2.1 Images' Specifications

It is utmost significant to know the specifications of an image acquired mainly by satellites, and also by other platforms. These specifications are usually primarily determined before selecting the image type. This depends on the purpose of the study and the spatial and temporal dimensions needed.

In this respect, there are many specifications characterize the images. The most important of these specification are:

1. Spatial Resolution

It is simply the size of a "pixel" that recorded in a raster images. Where the pixel is a picture rectangular element. Thus, pixel size enables distinguishing features and then identifying the objects on Earth's surface from satellite images (or other shuttle). In this respect, the entire remote-sensing system is significant, including lens antennae, display, exposure, processing and many other specifications to render a well recognizable image.

Therefore, low-resolution images are characterized by large pixel size where they are not able to show objects as clearly as high-resolution ones do. In this view, there are a number of satellite images with different spatial resolution that range from tens of centimeters (e.g. Geo-Eye, 41 cm; Quick-bird, 61 cm) up to few of kilometers (AVHRR, 4 km).

2. Spectral Resolution

Spectral resolution is attributed to the band range or band width reported by the sensor. Therefore, it is the ability of a sensor define fine wavelength intervals. Hence, the finer the spectral resolution, the narrower the wavelength range for a particular channel or band. For example, a Landsat satellite image has seven bands, including several in the infrared spectrum, where the spectral resolution ranges between 0.7 and 2.1 μm. While, an Aster satellite image has 14 spectral bands where 3 of them are visible and 11 in the infrared range.

3. Revisit Time

It is also described as the temporal resolution of an image where it is referred to the temporal frequency of sampling by repeat imaging. Therefore, it is defined as the time required by the satellite to comeback over the same point on Earth's surface. This is in turn dependent on the orbitography of the platform or satellite on which the sensor is hosted. With most of the Earth observation satellites having quasi-polar orbits, this frequency of revisits also depends on the latitude of the area of study. Thus, it is significant remote sensing element for mapping and monitoring and change detection approaches.

The spectral resolution (or revisit time) on an image does not relate to its spatial or spectral resolution. For example, Landsat (30 m) and Aster (15 m) require 16 days to make one turn around the Earth, while Spot-6 (6 m) needs 26 day to make the same turn.

**Table 4.1** Common observatory satellites with their major specifications

| Satellite | Number of bands | Spatial resolution | Revisit time | Swath width (km) |
|---|---|---|---|---|
| Worldview-4 | 6 | 0.31 | 1.7 days | 13.1 × 13.1 |
| Geo-Eye | 5 | 0.50 m | 2.8 days | 15.2 × 15.2 |
| Quick-Bird | 5 | 0.61 m | 1–3 days | 16.5 × 16.5 |
| IKONOS | 5 | 0.82 m | 3 days | 11.3 × 11.3 |
| Rapid-Eye | 5 | 5 m | 5.5 days | 77 |
| Sentinel-2 | 12 | 10 m, 60 m | 10 days | 290 |
| SPOT-7 | 4 | 1.5 m | 26 days | 60 × 60 |
| Aster | 14 | 15 VNIR, 30 m SWIR, 90 m TIR. | 16 days | 60 × 60 |
| IRS 1D | 4 | 23 m | 5 days | 70 × 142 |
| Landsat 7 ETM+ | 8 | 30 m, 120 m thermal, 15 m pan | 16 days | 183 × 183 |
| MODIS | 36 | 250 m, 500 m, 1 km, 2 km, 4 km | Twice/day | 2030 × 1354 |

4. Swath Width

The area imaged on the Earth's surface, is described to as the swath. Thus, swath width represents the areal coverage that sensors can viewed and imaged, and therefore produced settled image size (one scene). Hence, in aerial photogrammetry, swath width mainly depends on flight altitude type; nevertheless, the altitude has no role in swath width for images retrieved by satellites where it depends on the specifications of the sensor.

Even though, it is not usually the case, but several satellite images with large swath width are found with lower spatial resolution; whereas swath width and image resolution have no relationship. For example, the swath width of Sentinel-1 images is 80 km × 80 km, while it is 60 km × 60 km for Aster and Spot-7 satellite images.

It is; therefore, obvious that the specifications of images, notably those acquired by satellites, are significant in the selection of an image for specific terrain analysis. Table 4.1 shows selective, and commonly used satellite images.

### *4.2.2 Satellite Image Processing*

Considering the optical and spectral specifications; however, the efficiency of these satellite images has been well demonstrated lately, especially in the assessment and study of many themes on the Earth's surface including monitoring approaches, natural hazards, change detection and many other applications.

As an example, the observation of terrain surface from space by Ikonos satellite images enables viewing objects of approximately 0.82 m × 0.82 m area, which is equivalent to watch these object from space at about 100 m elevation. This virtually means that mankind can fly over any area and look down and distinguish all objects exceed 0.67 $m^2$, but this observation can be each three days only. Therefore, more than one type of satellite images are used for several goals including dates overlapping, utilizing from the spatial resolution of one image to another, etc.

It is essential to know how to select the most appropriate satellite images including mainly the spatial and spectral resolution along with acquiring periods (revisit time). Another significant factor is the expertise and capability to analyze these images with proper computerized devices. In addition, satellite images have various spectral and electronic properties, such as the number of bands, their wavelength and spectrum ranges (e.g. thermal band, microwave, optical band, etc.). Thus, it must be made clear all these details should be primarily known.

The availability of satellite images is often a constraint. However, it was a smart decision taken by the Kingdom of Saudi Arabia when it adopted successful policy towards space techniques, and thus the Space Research Institute has been established at the King Abdulaziz City for Science and Technology (KACST). In this respect, KACST has introduced several facilities, and at different time periods, to the author to perform studies using space tools, specifically the satellite images of high resolution.

The processing of satellite images starts when the digital (electronic) raw data as well as the appropriate software become ready. In this respect there are several software types used, the most commonly ones are:

- ENVI Image Analysis: Produced by: IBM. Colorado, USA.
- ERDAS Imagine: Produced by: Lucia, Georgia, USA.
- PCI Geomatics: Developer of Geomatica, Toronto, Canada.
- ILWIS: Produced by ITC, Enschede, Netherlands.

Processing of satellite images with specialized software follows defined technical approaches. These imply the pre-processing steps and image analysis and classification. They can be concluded as follows:

1. Pre-processing: This is the primary phase applied to prepare the images for further analysis and classification. Therefore, satellite images cannot be analysis before applying the following processes on the used soft water.

    - Image sub-setting, is the process by which the area of interest (AOI) is extracted from the entire images scene (example Fig. 4.1). This in order to avoid the slow-down the work and any demanding on computing resources.
    - Atmospheric correction, is applied for noise removal and to identify true surface reflectance by removing atmospheric effects from satellite images.
    - Geometric correction, is the noise and sun-angle correction that always result in images displacement due to the high altitude of shuttle-bearing sensor. Thus, registration is applied using "rubber sheet" transformation which warps the image on defined points.
    - Geo-referencing, is the rectification of an image for the assignment of geographic location, scale, and alignment to a file, and more specifically it is

**Fig. 4.1** Example of satellite images sub-setting

performed on raster and vector data to interrelate internal coordinate system to associate something with locations in physical space.

- Mosaicking, is almost a reverse process for sub-setting, but in this case, multiple images on different separate scenes are correlated together in order to have a unified scene for the AOI (e.g. watershed, cadastral unit, etc.).

2. Image analysis and classification: application: When the image is prepared following the previous pre-processing steps, the next phase is the image analysis and classification (i.e. image processing) in order to identify the existing objects, calculating dimensions, monitoring and many other purposes required. This can follow two major methods:

- Observation enhancement and detection, by which several digital and spectral applications are applied in the used software aiming at observing and identifying objects clearly as much as the resolution of the images allows. These application are mostly tentative and their use is dependent on the skill of the analyzer.
  The most common of these applications are the: color slicing, edge detection, directional filtering, enhancement, interactive stretching, contrasting and sharpness. In addition, combination is also performed where single band and multi-band enhancement are carried out by interrelating each three bands as one set.
  Image classification, uses "Classifier" on the software and it involves grouping the image pixel values into indicative categories. There are main types of this classification where the first two are the most applied (GIS Geography 2020a). These are: (1) Unsupervised classification where it groups pixels into "clusters" based on their properties, and the each cluster can be tentatively attributed to land class, (2) Supervised classification where the analyzer selects representative samples for each land class, and then the software uses these "training

sites" and applies them to the entire image, (3) Object-based image classification groups pixels into representative vector shapes with size and geometry. It is not therefore similar to the supervised and unsupervised classification which are pixel-based and create square pixels and each pixel has a class.

## 4.3 Geo-information System

Geo-information system (or Geographic Information System-GIS) is a computer system for extracting, storing, checking, drawing, displaying the geo-spatial data. It is capable to show several aspects of geo-spatial data on one map, such as streets, buildings, and vegetation (Fig. 4.2). This enables people to more easily see, analyze, and understand patterns and relationships (NGS 2020).

The GIS technology perform digital information, for which various digitized data illustration and methods are used, such as the digitization of data, where hard copy maps or survey plans are transferred into digital medium through the use of a computer-aided design and geo-referencing capabilities. Recently, GIS has been involved within the board of several institutions (e.g. ministries, research centers, universities, etc.) as a managing tool in urban planning, transportation, water quality monitoring, etc.

There are several software used for GIS applications. The most commonly used is ESRI (Environmental System Research Institute, Redlands, USA). It is widely used where it broadly implies *Arc-GIS,* as the principal Geo-information system tool extended on ITS computers, and is installed in UNIX and Networked PC devices.

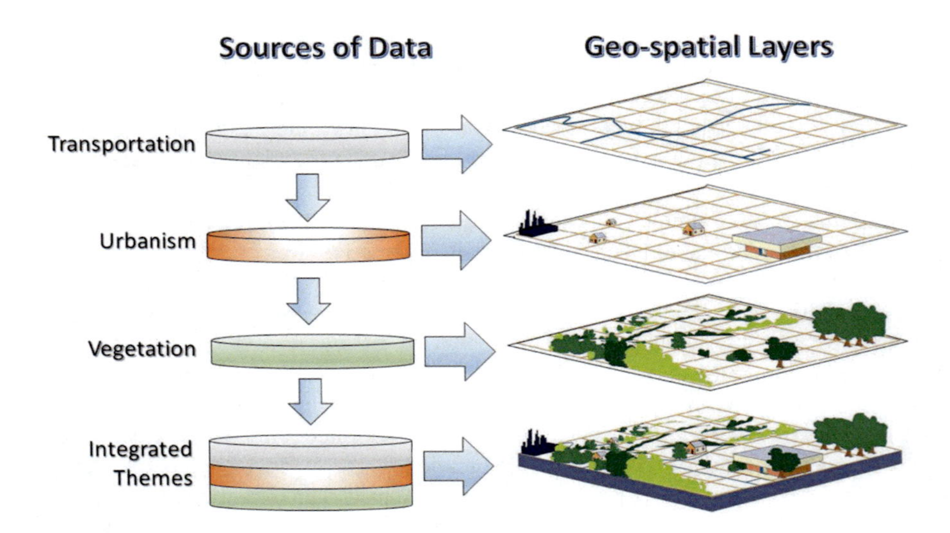


**Fig. 4.2** Example showing the integration of different geo-spatial layers in the GIS system

*Arc-GIS*, as a software, is utilized to generate, display and analyze geo-spatial data. It includes three digital components:

- *Arc-Map* which enables visualizing spatial data, performing spatial analysis and drawing maps.
- *Arc-Catalog* is a tool for browsing and exploring spatial data, as well as viewing a creating metadata and managing spatial data
- *Arc-Toolbox* is an interface for accessing the data conversion and analysis function the come from *Arc-GIS*.

## 4.4 Topographic Map

Topographic maps are usually generated as a first base document up on which other thematic maps and illustrations needed are plotted. This has been previously mentioned in this chapter. Thus, for SLM approaches planned by the author for NEOM Region, a unified topographic map was one of the first products, because no topographic map has been done specifically for NEOM Region before.

### *4.4.1 Map Production*

The available topographic maps are present as separate map sheets (hard copies) within the context of the produced topographic maps for the entire Saudi Arabia. Which have been done by the Aerial Survey Department under the supervision of the Ministry of Petroleum and Mineral Resources (MoPMR) in (1970).

Three topographic map sheets at scale 1:250.000 were used to cover the entire NEOM Region, and they belong to quadrangles of:

- Haqel map (sheet: 29 A).
- Al-Bada'a map (sheet: 28 A).
- Al-Muieleh map (sheet: 27 A).

Due to its significance as one of the basic documents, the author identified the digital forms of the border limits for NEOM Region which was initially determined by the concerned governmental organization; and therefore, the terrain surface data for this region were extracted. This has been based on the existing data and information about the area of study, and then updating approaches were performed for data and information in consistency with terrain components in 2019 when NEOM Region topographic maps has been produced by the author.

The production of the topographic map of NEOM Region has been achieved (Fig. 4.3), as the first topographic document of its type for this region. This was based on the database of the old topographic maps in integration with data used from the SRTM (Shuttle Radar Topography Mission) satellite images which severed in creating the DEM and the contour lines. Of course, this has been accompanied with

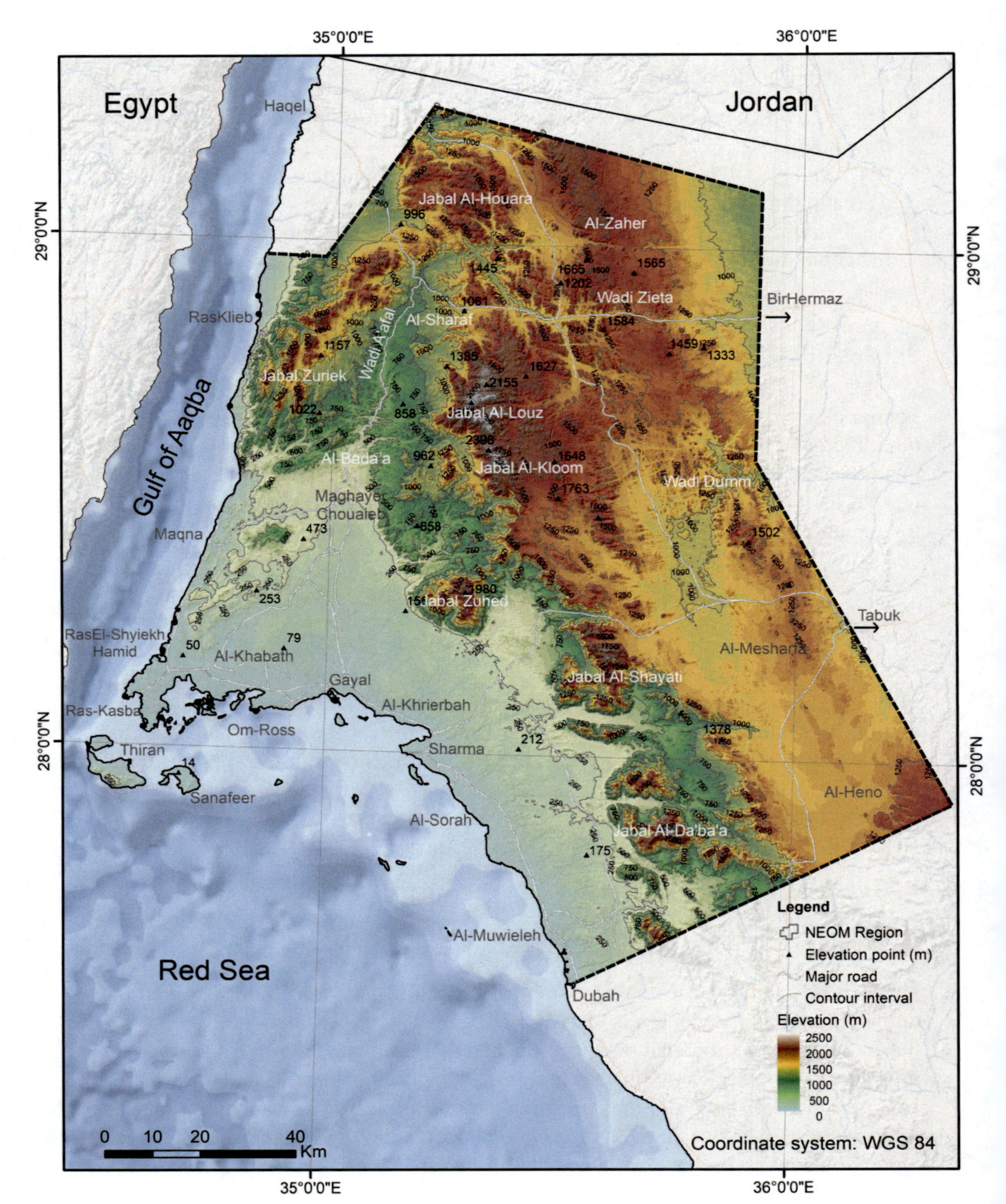


**Fig. 4.3** Updated and enhanced topographic map for NEOM Region

field survey. Hence, the produced map followed the coordinate system: WGS 84 which comprises a reference ellipsoid, a standard coordinate system, altitude data and a geoid.

### 4.4.2 Map Advantages

The production advantages of the new produced topographic map for NEOM Region implies the following:

1. One-sheet map dedicated for NEOM Region where it can be plotted at different paper size with high resolution.
2. Large-scale maps (1:200.000), instead of 1:250.000 scale of the previous available maps.
3. It contains DEM background with color scale bare.
4. New and updated names replaced the old ones.
5. New terrestrial (e.g. roads, urban settlements, etc.) and marine (e.g. ecosystem) components have been added.

The technical advantages of the new produced topographic map for NEOM Region implies the following:

1. Data of the produced map are digital in GIS format (i.e. shape-files) and composed of different geo-spatial layers. This enables merging different thematic maps (as layers), as well as adding/or eliminating new themes or data whenever it is needed.
2. It is capable form the produced map to calculate the area of any zone with define elevation (e.g. vegetation according to altitude, etc.).
3. It enables calculating the relationship between specific themes with respect to altitude (e.g. snow line/altitude, etc.).
4. The produced topographic map is very significant for urban planning approaches.
5. The digital boundary of NEOM Region, as a GIS shape-file can facilitate using it for further applications since the area boundary has been newly determined.
6. For NEOM Region, the produced maps was used as the base maps for other thematic maps done also for the same region. This will be illustrated in the next sections.

## 4.5 Geology Map

It is necessary to prepare an updated and creditable geologic map for NOEM Region upon which several related works (e.g. studies, projects, etc.) will be applied for this region which is supposed to be a global economic hub. Therefore, such a unified map can be used even by individuals and not only by geoscientist, such as by tourists, travelers, etc.

As previously mentioned, the available produced geological maps have been done by the MoPMR when different regions of the Saudi Kingdom were mapped by different experts and at different dates. Therefore, three quadrangles maps sheets (i.e. Haqel by Rowaihy (1985); Al-Bada'a by Clark (1987); Al-Muieleh Davies and Grainger (1985) cover the geology of Neom Region.

The available maps, hard copies with 1:250.000 scale, compose all essential geologic features including mainly the lithological distribution and geologic structures. These features are:

- The geographic illustration for different rock types (i.e. lithologies).
- Major geologic structures (e.g. faults, anticlines, etc.).
- Economic geology attributes by each rock formation and the related sites.
- Selective geologic cross-sections with dimensional elements.
- Detailed legend including rock sequences and rock formation description, plus all symbols for the geologic structures.
- Detailed explanatory notes (booklets) were elaborated for each maps sheet.

### *4.5.1 Map Production*

Even though the obtained geological maps, by MoPMR, have been done since more than 35 years ago, yet these maps are still used since they contain valuable geologic data and information, as well as considering the fact that the geological features do not change over such time periods. This was helpful and supportive in order to generate a unified and enhanced geologic map dedicated specially for NEOM Region (Fig. 4.4).

Several steps were applied in order to create the new geological map for NEOM Region where remote sensing and GIS techniques were principally adopted. These steps can be summarized as follows:

1. The plotted polygons (i.e. for different lithologies) and linear features (i.e. for faults and other geologic structures) on the old maps were primarily digitalized using Arc-Map package included in the Arc-GIS, and shape-files were imitated.
2. Satellite images were used, and this included the Landsat 7 ETM$^+$ (30 m, 60 m TIR), Sentinel-2 (10, 20 m SWNI), Spot-7 (1.5 m, 10 NIR). In this respect, ERDAS-Imagine 2018 was used. These images were subjected to all necessary pre-processing approaches including sub-setting, mosaicking, radiometric corrections, etc.
3. Therefore, all required digital and optical advantages were applied including (for example): edge detection, directional filtering, contrasting and sharpness, etc.
4. Raster image files were generated for the following for modifying the geologic boundaries for different rock formations using visible and TIR bands in Landsat 7 ETM$^+$, and SWNI in Sentinel-2, and sometime WV-2 images were also used. In addition, single-band and multi-band enhancements were applied by integrating each of the three bands as one set where the combination of bands 2, 5 and 3 in

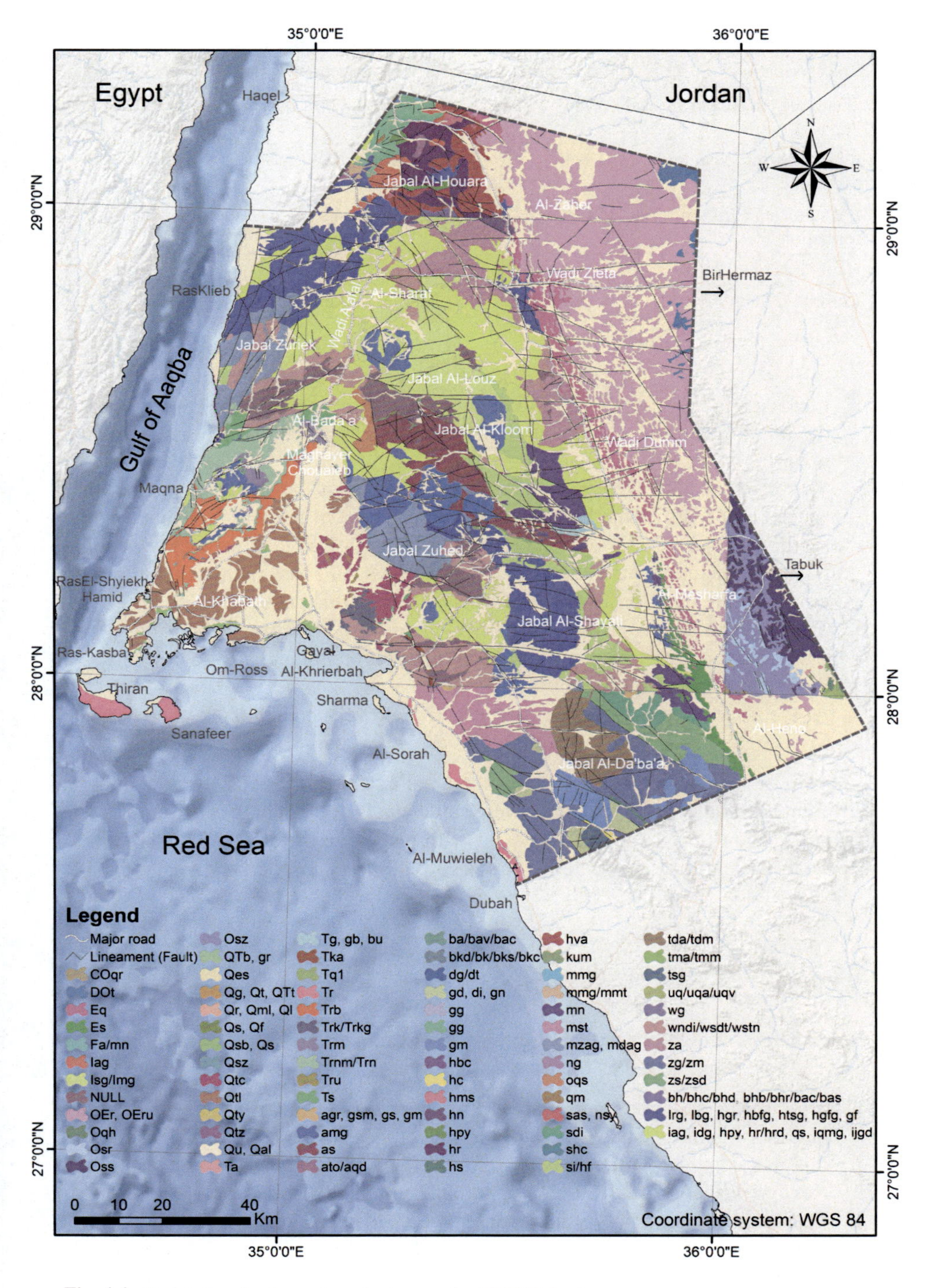


**Fig. 4.4** Updated and enhanced geologic map for NEOM Region

**Table 4.2** Areas of different rock formations in NEOM Region

| Rock formation | Area (km$^2$) | %[a] | Rock formation | Area | %[a] |
|---|---|---|---|---|---|
| Quaternary sand and alluvial deposits | 5480 | 21 | Andesite and lava (Proterozoic) | 662 | 22.5 |
| Ram and Umm Shamm sandstone | 1814 | 7 | Atiyah Monzonite | 506 | 2 |
| Ifal suite and Hinshan formation | 1474 | 6 | Lawaz Complex | 431 | 1.5 |
| Haql Suite | 953 | 4 | Midyan Suite | 418 | 1.5 |
| Muwaylih Suite | 680 | 3 | Hegaf Formation | 316 | 1 |

[a]Percentage of the total NEOM Region

Landsat images to reach the most clear and indicative observation for recognized the lithological extent and boundaries which were plotted on the new geologic map (Fig. 4.4).

5. Vector image files were established in order to plot the linear features with a special emphasis on the fault structures which are observed as "lineaments" the satellite images. Therefore, the recognized fault, other than the ones exited on old geologic maps, were added to the map (Fig. 4.4). These fault were performed on a separate map for further applications later in this document.
6. The rock formations on the old geologic maps were also investigated and specifically the ordering of these formations, their chronological attributes as well as the contradictory nomenclatures were fixed.
7. Al the above step were accompanied with field verification and reconnaissance to assure the extent of some lithologies as well as the recognized faults.

From the produced geologic maps, the areal extent of the rock formations with different lithological characteristics were calculated since this map is in digital form. Table 4.2 shows the largest ten areas of with respect to different rock formations. Hence, it is clear that the Quaternary rock formations occupies the largest area.

## 4.5.2 Map Advantages

Nevertheless, the author added several new advantage on the produced maps and these can be summarized as follows:

1. Similarly to the topographic map, the produced geologic map is obtained digitally in GIS system (i.e. shape-files) where several geological layers are present.
2. It is also a one-sheet map especially done for NEOM Region with high resolution.
3. Newly existed terrain features (e.g. urban settlements, roads, etc.) were plotted.
4. There was a contradictory included between different map sheets (obtained by MoPMR), notably in the nomenclatures of rock formations and their symbols. This has been corrected and mangled into unified manner.

5. New faults were plotted on the produced map where satellite images were the tool used for this purpose. Therefore, the existing faults on old maps remained, and their alignments were sometimes modified; and thus the new extracted faults were added to the produced map.

## 4.6 Landforms Map

Landform is a define geomorphic feature on terrain surface that exist in different dimensions starting from ranging large terrain features such as plains, plateaus and mountains to relatively smaller terrain features such as hills, valleys and alluvial fans. Thus, the term landform differs from geomorphology in that it often points out to the physical processes occur on it.

Landforms are created over different time periods either as flash events or they may take many decades to form. They are resulted by several processes where they can be classified principally into two main types: (1) build-up of surface features either by physical/or man-made influencers (2) destructive or the breakdown of land surface to form new features. Thus, the majority of landform processes can include, in a broad sense, crustal deformation, volcanics, sedimentation, weathering and erosion.

Usually landform maps are prepared for projects using geo-spatial information. This is well known when there are active surficial processes occurred/still occurring, as well as when geomorphological maps are not available. Therefore, it is integral scientific document necessary for studying SLM such as in the case of NEOM Region.

As previously mentioned that NEOM Region has four diverse topographic units and these units, along with the complicated geology, make the region vulnerable for surficial processes whether constructive (i.e. build up) or destructive processes. In addition, landform map can substitute land cover/use map in an area like NEOM. This is because the human activities in NEOM are still negligible and the geo-spatial data in landform map can fulfill the scope of characterizing the topographic processes of the region.

The available landform maps done by the Department of Land Management at the Ministry of Agriculture and Water (MoAW 1980). They imply similar quadrangles of the topographic and geologic map sheet which are: Haqel map (sheet: 29 A), Al-Bada'a map (sheet: 28 A) and Al-Muieleh map (sheet: 27 A).

The available landform maps have the following specifications:

- The available landform maps obtained by MoA are hard copies and at scale 1:500.000.
- Detailed explanation and discussion are available for each of these maps. This is not only for the defined landform polygons (i.e. geographic classes) but also for selected sites within these polygons.
- Detailed legend is available and they appear different from one map to another. The legend includes landform classes with a special focus on the agricultural lands.

- Complementary terrain features (e.g. roads, urban areas, etc.) are plotted on these maps, but not the contour lines.

### *4.6.1 Map Production*

The production of landform map of NEOM Region was based on available landform maps obtained by the MoA (1980). However, the production of the new landform map for NEOM Region followed these technical steps:

1. Similarly for the produced geologic map, the polygons (i.e. landform classes) on the old maps were initially digitalized using Arc-Map package, and therefore, shape-files were generated. This was limited to NEOM Region border.
2. Satellite images were also used here, and this implies the following images: WV-2 (0.46 m B&W); Spot-7 (1.5 m, 10 NIR). Hence, ERDAS-Imagine 2018 was also used. All required pre-processing approaches including sub-setting, mosaicking, corrections, etc.
3. Digital and optical advantages available on ERDAS Imagine 2018 were utilized such as: band combination (mainly band order: 4, 3 and 2), edge detection, directional filtering, color slicing, etc. This enabled identifying the boundaries of landform classes as well as correcting the geographic extent of some polygons.
4. Field survey has been achieved to verify the reliability of data and information obtained by satellite images.

Therefore, a unified and updated landform maps for NEOM Region has been produced as shown Fig. 4.5.

### *4.6.2 Map Advantages*

The produced landform map for NEOM Region has several advantages makes it more applicable for use with better and easier way than the available maps of landform. These advantages are:

1. The produced map is one-sheet map with detailed legend.
2. The map is in digital form (GIS files), with different shape-file themes, and this enables applying several systematic integration with other themes when it is needed. In addition, all dimensions can be calculated, such as the area of each class (Table 4.3).
3. The new map has larger scale than the old maps where it can be illustrated clearly on 1:250.000 scale.

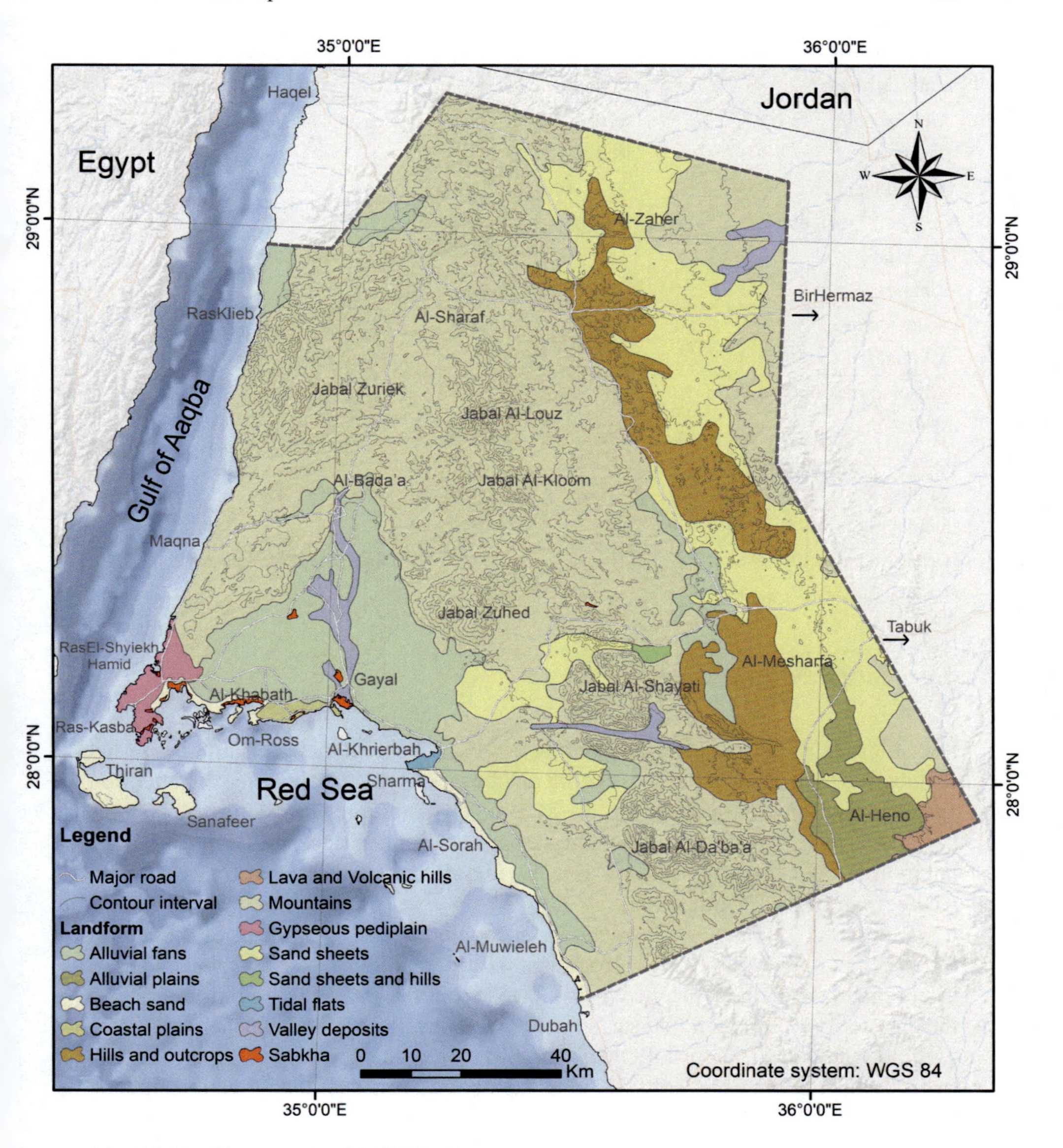


**Fig. 4.5** Landform map for NEOM Region

4. The new landform map contains the contour lines as a significant terrain feature enabling to figure out the geographic distribution of each landform type with respect to altitude.
5. New terrain components (e.g. roads, urban areas, etc.) are plotted on the new map.

Table 4.3 Areas of different landform and their geomorphologic attribute in NEOM Region

| Landform type | Area ($km^2$) | %[a] | Geomorphologic attribute |
|---|---|---|---|
| Alluvial plain | 816 | 3.07 | Coastal zone |
| Beach sand | 324 | 1.22 | |
| Coastal plain | 367 | 1.38 | |
| Gypseous pediplain | 181 | 0.68 | |
| Tidal flats | 84 | 0.32 | |
| Sabkha | 78 | 0.29 | |
| Valley deposits | 756 | 2.85 | Valley |
| Alluvial fans | 3368 | 12.70 | Mountain ridges |
| Hills and rock outcrops | 2302 | 8.68 | |
| Mountains | 14,015 | 52.88 | |
| Laval and volcanic hills | 182 | 0.68 | Plateau |
| Sand sheets | 79 | 0.30 | |
| Sand sheets and hill rocks | 3845 | 14.51 | |
| Urban and agricultural area | 103 | 0.40 | Distributed in all above zones |

[a]Percentage of the total NEOM Region

## 4.7 Slope Map

Even thought, slope is an essential geomorphological aspect that acts on/or below the surface and then influencing the regolith and bed rock, yet it is always given attention and it is often investigated separately as a major feature of Earth's surface. This is because surface steepness is important in shaping terrain surfaces and controlling the geomorphological processes such as gully and rill erosion, rain-wash, sheet-wash, weathering, mass movement, piping, etc.

Therefore, slope is a principal factor accounted in the assessment of on natural hazards (i.e. mainly floods and terrain instability), and thus it must be considered in urban planning methods. Hence, it is viewed from the responding point of view to the disaster risk reduction that sustainable urban development (El Kechebour 2015).

In the view of SLM for NEOM Region, it would be necessary to generate slope map for this region, especially this map can be used for several applications related to surficial processes, distribution of urban settlements and for risk management. However, there is no slope maps done for NEOM Region before, and this map will be the first of its type where new advanced techniques are used.

To producing the slope map for NEOM Region, SRTM (Shuttle Radar Topography Mission) satellite images were used in order to initially generate the Global Digital Elevation Model (SRTM DEM) as a systematic tool to extract slope for the area of study.

SRTM DEM becomes an effective tool to generate slopes of any terrain. Even though DEMs have been mentioned previously in this chapter; however, more highlights are given to this technique on slope production.

Digital Elevation Models have been recently adopted in many studies on terrain behavior. These models are generally achieved using GIS systems. This can be done from digital topographic maps or from stereoscopic satellite images where Triangulated Irregulated Networks (TINs) are constructed for this purpose.

Recently in 2014, SRTM (Shuttle Radar Topography Mission) satellite products were extended by the National Aeronautics and Space Administration (NASA). It was used to initially to generate SRTM DEM as a systematic tool to extract slope for the area of study. This tool (SRTM) made of radar interferometry, i.e. two radar images are taken from slightly different locations to produce digital topographic data for 80% of the Earth's land surface, with data points located every 1-arc-second (approximately 30 m) on a latitude/longitude grid (USGS 2020). Therefore, SRTM DEM can be downloaded, and it is available at GIS Geography (2020b). Thus, the downloaded SRTM DEM cab be produced by different orientation, colors and from different view angle as showing Fig. 4.6.

Therefore, the following points describes the steps followed to generate DEM for NEOM Region wherefrom slope map was extracted using GIS system.

1. The SRTM DEM will be processed by Arc-GIS software, and specifically in the Toolbox system.
2. The navigation will be through the Toolbox-then Spatial Analyst tools-Surface-Slope.
3. Raster files are then opened and the location of the output raster is specified and eventually the output measurement will be selected as well.

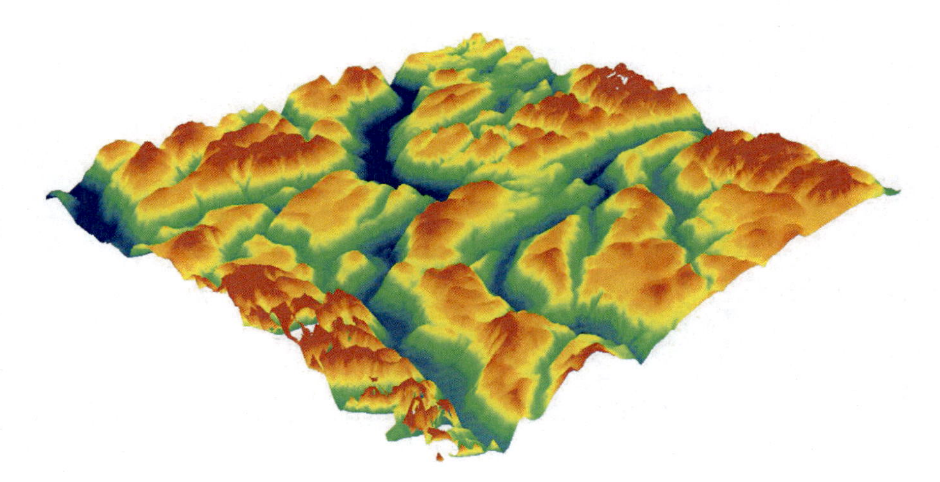

**Fig. 4.6** Example of generated SRTM DEM (GIS Geography 2020b)

Therefore, the slope map was created with five classes were identified with specific colors. These classes range as follows: <5°, 5°–10°, 10°–20°, 20°–30° and >30°. It is; therefore, the first slope map for NEOM Region (Fig. 4.7).

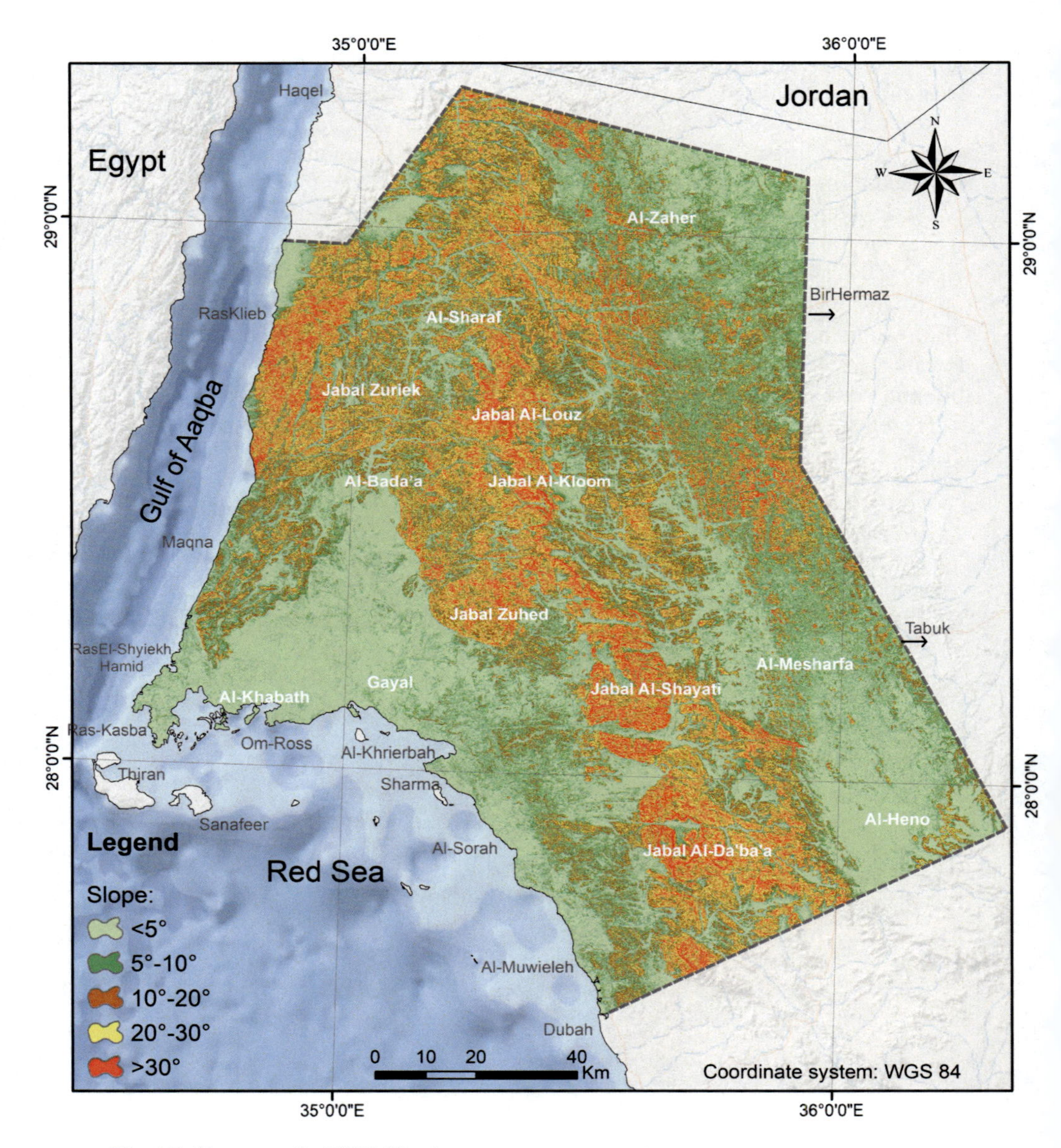


**Fig. 4.7** Slope map for NEOM Region

It is obvious that the most dominant slope in NEOM is below 5° where it occupies about 10,940 km$^2$. Then the rest slopes are distributed as: 4956, 5274, 3318 and 2012 km$^2$ for slopes: 5°–10°, 10°–20°, 20°–30° and >10°; respectively.

## References

Alrajhi, M. (2005). Overview over the activities of the general directorate for surveying and mapping and its future perspectives. In *8th UN Regional Cartographic Conference for the Americas* (5 pp). NY, June 27–July 1, 2005.

Clark, M. (1987). *Geologic map of Al-Bada'a quadrangle, A-28; (1:250.000).* Ministry of Petroleum and Mineral Resources.

Davies, F., & Grainger, D. (1985). *Geologic map of Al-Muieleh quadrangle, A-27; (1:250.000).* Ministry of Petroleum and Mineral Resources.

Dixon-Gough, R. W. (1994). Geographical information management: the way forward for remote sensing. *Geodetical Information Magazine, 8*(8), 68–74.

El Kechebour, B. (2015). Relation between stability of slope and the urban density: Case study. *Procedia Engineering, 114,* 824–831. https://doi.org/10.1016/j.proeng.2015.08.034.

GIS Geography. (2020a). *Image classification techniques in remote sensing*. Available at https://gisgeography.com/image-classification-techniques-remote-sensing/.

GIS Geography. (2020b). *Digital elevation model data sources.* Available at https://gisgeography.com/free-global-dem-data-sources/.

MoAW (Ministry of Agriculture and Water). (1980). *Landform maps.* Riyadh, KSA: Department of Land Management.

MoPMR (Ministry of Petroleum and Mineral Resources). (1970). *Topographic maps of Saudi Arabia, 1:250.000.* Riyadh, KSA: Aerial Survey Department.

NGS (National Geographic Society). (2020). *GIS (Geographic information system).* Resource Library. Encyclopedic Entry. Available at https://www.nationalgeographic.org/encyclopedia/.

Rowailhy, M. (1985). *Geologic map of Haqel quadrangle, A-29; (1:250.000).* Ministry of Petroleum and Mineral Resources.

Saudi Arabia (Jeddah-Yanbua) (1969). KSA: Department of Hydrology, King Abdulaziz City for Science and Technology.

USGS (United States Geology Survey). (2020). *USGS EROS archive—Digital elevation—SRTM Mission summary.* Available at https://www.usgs.gov/centers/eros/science/.

# Chapter 5
# Drainage Basins

**Abstract** Drainage basins are significant terrain features which are usually analyzed in many studies and engineering applications. They represent essential components of surficial processes associated with natural hazards and water storage. This must be given attention when drainage basins are well developed like the case in the Arabian Peninsula where valleys can be clearly observed from space shuttles and they are the most dominant features as much as the deserts. Studies on drainage basins have been widely done in Saudi Arabia, and the number of these studies has been increased when flood events recurrently took place and resulted severe damages. Recently, most studies obtained on drainage basins use new advanced techniques of remote sensing and geo-information systems. There are new space shuttles which are able to retrieve stereoscopic and radar images, and then enable generating the three dimensional view upon which streams can be projected and their catchments can be therefore identifies. NEOM Region, as a bare area, lacks to drainage basin map and no geometric or morphometric calculations have been achieved for this region which will include several developing projects. Therefore, drainage basins must be studied in-depth and all related calculations should be prepared as an objective of this chapter. The outcomes will be very supportive data and information used in SLM application, notably in the themes of flood assessment and surface water harvesting practices.

**Keywords** Stream · Water divide · Meandering ratio · Drainage morphometry · SRTM

## 5.1 Introduction

The Arabic word "Wadi" means valley, which represents two watercourse-facing terrain bodies along watercourses (i.e. also streams or drainages), whereas a drainage basin includes wadis. In the Kingdom of Saudi Arabia wadis are significant agricultural lands where surface water are often exist whether beyond the executed dams, in the constructed canals or as groundwater in the alluvial deposits along valley banks. Thus, watercourses have resulted in recent floods and torrents in many regions of Saudi Arabia. These geomorphologic features are likely sufficient to underline the



M. M. Al Saud, *Sustainable Land Management for NEOM Region*,
https://doi.org/10.1007/978-3-030-57631-8_5

importance of drainage basins and warrant detailed studies in order to preserve the water wealth and avoid the occurrence of any natural disasters resulting from the flow of this water (Al Saud 2018b).

More importance should probably be attached to those valleys that are adjacent to/nearby that actually cut across major cities and towns. Therefore, the study of drainage basins are often included with the hydrology of any area or while preparing the thematic maps needed for father assessment. Whereas, in this document the study of drainage basins has been put as a separate chapter due to its significance in SLM.

There are 14 mega basins exist in Saudi Arabia with a total aggregate length of about 45,000 km; as well as there are hundreds of large-scale basins (few thousands of square kilometers) with their streams are also exist. These basins have wadis with diverse dimensional aspects and with different orientations. Besides, the hydrology of the western Saudi Arabia, including wadis, is considered as a distinguished and unique hydrological features pertaining to the recent and ancient aspects of the hydrological processes (Al Saud 2011). These wadis span from the mountainous regions in the east and then outlet their seasonal water in the Red Sea.

No doubt, the study of drainage basins should be always applied prior and study concerned with geo-spatial data analysis. Thus, to study the drainage basins, stream maps and the surrounding elevations must be primarily prepared, and if these maps are digitally produced, this enables applying different required calculations whether for the dimensional aspect of the basin and its boundary or for the streams (with different dimensions) included with this basin.

## 5.2 Cartography of Drainage Basins

Drainage basins cartography and analysis has significant contribution in many applications, certainly in a region like Saudi Arabia where the produced drainage basin maps and their calculations are used in many studies such as in flood assessment, selecting site for dams construction and other water harvesting approached, as well as in mass movement and the related erosion processes.

Drainage basins of the Kingdom of Saudi Arabia have been tackled in several applied studies (examples: Aawari 2005; Qari 2009; Subyani et al. 2009) where the author elaborated a big number of these studies and covered approximately 25% of the drainage basins in the entire Kingdom (examples: Al Saud 2007; Al Saud 2012; Al Saud 2018a, b). Nevertheless, except some studies done on floods in Tabuk city (example: Al-Momani and Shawaqfah 2013; Abdelkarim et al. 2019), no studies done yet on the drainage basins for the western part of Tabuk Province, including NEOM Region.

Cartography of drainage basins was manually obtained from the tracing of streams on topographic maps. This conventional method remained for long time until the new remote sensing and geo-information techniques have been raised. Therefore, a wide range of methods have been done to facilitate extracting the streams and their catchments boundary. In this respect, again the GDEM products along with the Arc-GIS system has the principal role in the extraction of drainage basins and in calculating all required measurements needed for each theme whether for floods, dams location, etc.

Similar methods previously used, by the author, in extracting drainage basins for Riyadh Region, were used for NEOM Region which were used using SRTM DEM and Arc-GIS basin. Thus, channel initiation nodes were primarily identified from the digital raster sources using SRTM DEM with a 30 m spatial resolution. This was based on the recognition of the incipient flow direction where the used SRTM helped identifying the water divide points and then mapping watersheds. For this purpose, D8 flow direction algorithm available in the Spatial Analyst extension of Arc-Map was used with because it contains a hydrologic toolset by which depressions are deemed drainage areas.

The following summarizes the step of stream network extraction from DEM:

1. A "Fill" tool is primarily applied to correct any imperfections in the DEM, and the artificial depressions were filled before extracting flow direction.
2. When the value of each cell gives a code for the direction of water flow; therefore, the matrix of flow direction can be generated, and the flow accumulation can be computed.
3. On the generated raster file, each pixel will assigned for a value that corresponds to the number of adjacent pixels that drain towards it.
4. A critical level (CL) for the flow accumulation must be identified to produce stream, where CL represents the number of pixels that indicate whether the pixel belong, or not to the stream course. Thus, CL values are selected by considering the spatial resolution of the DEM.
5. The lower CL values generates more stream branches and vice versa.
6. "Raster Calculator" tool in the Arc-GIS is used to select the values with flow accumulation larger than the CL.
7. The hydrologic modeling tools can determine the required correlation, and most importantly the flow direction and flow accumulation as well as Strahler ordering.

The steps of stream network extraction from DEM, as adapted from Monteiro et al. (2018) is shown in Fig. 5.1. Thus, the applied steps were done to find the stream networks and the related reaches, and then the flow direction from the upstream immediately to downstream areas. The catchment boundaries (basin boundary) is determined using the generated streams and the elevation points around each cluster of streams.

For each output basin and its streams, attribute tables with data packages were generated. This enables applying different morphometric and geometric formula based on mathematical and statistical calculations.

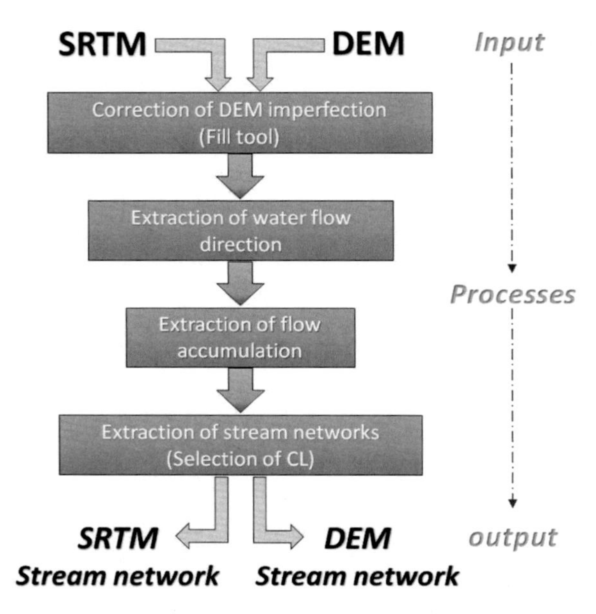


**Fig. 5.1** Flow chart showing the stream network extraction from DEM. Adapted from Monteiro et al. (2018)

## 5.3 Drainage Basins of NEOM Region

Following the methodology for drainage basin extraction; therefore, drainage basins of NEOM Region were generated for the first time. The extracted drainage basins include the stream networks and the outer boundary where these streams are included (i.e. watersheds). Hence, 45 drainage basins, with different dimensional aspects, have been recognized within the limits of NEOM Region. Hence, the watersheds of these basins are shown in Fig. 5.2.

There are 2 drainage basins originated from outside NEOM Region (basins number 32 and 42), but they pass in the region and outlet at its coast and then they can be described as inbound drainage basins (IDBs).

Besides, there are 6 drainage basins originated from NEOM Region (basins number 1, 2, 3, 4, 8 and 43), but they outlet outside its boundary and they can be described as outbound drainage basins (ODBs). In addition, 33 land areas located along the coastline where they join between the coastal drainage basins, but they are not considered as catchment areas because they neither encompass the characteristics of the catchment nor define primary watercourse, and thus they are described as undefined basins (UBs).

The principal components of the identified drainage basins (i.e. watershed and streams) are characterized by diverse dimensions, catchment shapes and stream orientations which directly reflect the beneath geology as well as the geomorphology of the area.

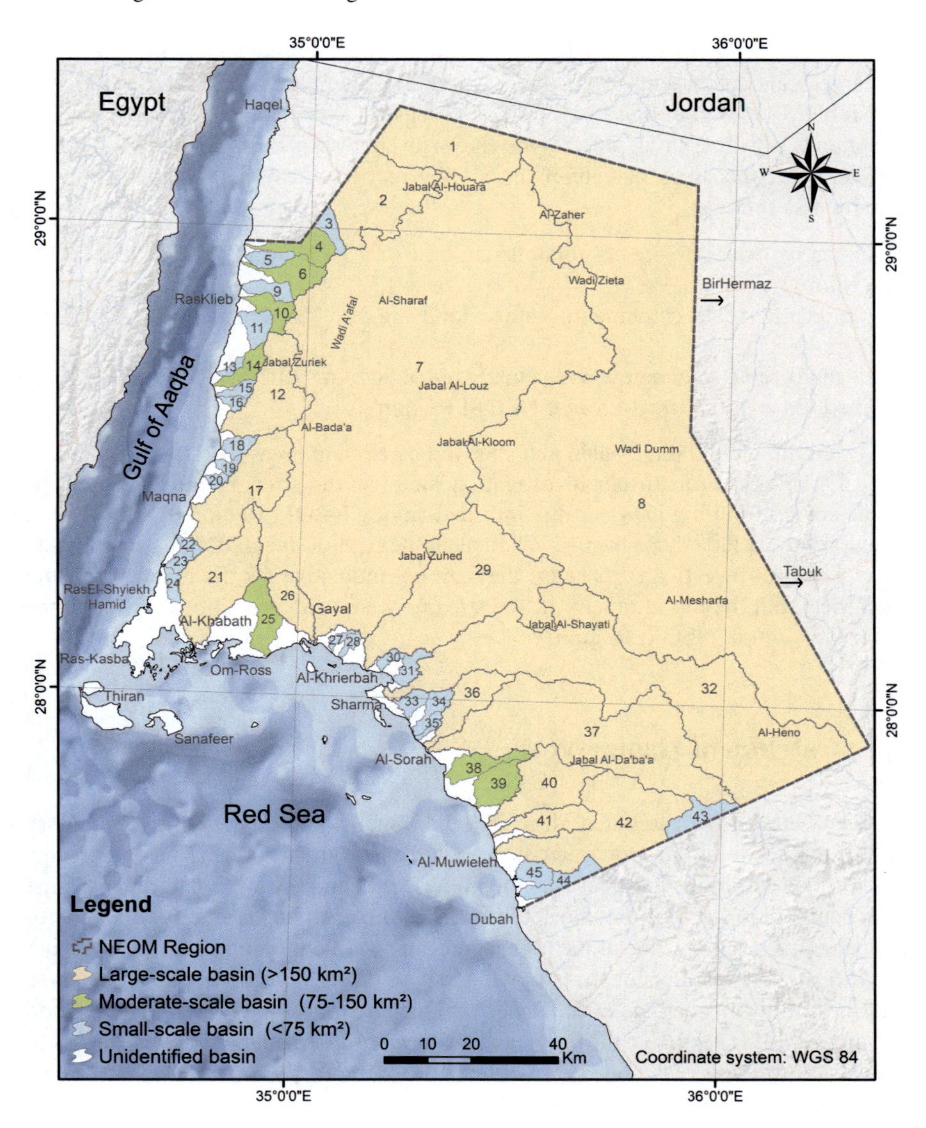


**Fig. 5.2** The catchments of the drainage basins of NEOM Region

Hence, 15 of the drainage basins have considerable catchment areas, and then they are large-scale catchments (>150 km$^2$, where 2 of them are of the ODBs (Fig. 5.2 and Table 5.1). While, there are 7 drainage basins with relatively moderate-scale catchments (75–150 km$^2$), plus 23 ones are with small-scale dimensions (<75 km$^2$).

Therefore, the areas of catchments of the recognized drainage basins can be classified as follows:

- 15 large-scale catchments totaling an area of 23,737 km$^2$ (89.5% of NEOM Region).
- 7 moderate-scale catchments with a total area of 739 km$^2$ (2.8% of NEOM Region).
- 23 small-scale catchments with a total area of 847 km$^2$ (3.2% of NEOM Region).
- The UBs = 1177 km$^2$ (4.5% of NEOM Region).

For creditable drainage basin analysis, the catchment areas of IDBs were added to NEOM Region for further investigation, because the streams with these catchments are contributing in water flux intruded among NEOM Region; therefore, it is necessary to include these basins. Whereas in the case of the ODBs, the catchments areas were not investigated because they are not indicative for the geometry of the catchment, but the morphometry of the streams were analyzed since they influence water flow within NEOM Region.

## 5.4 Catchment Geometry

The geometry of the catchment describes the dimensional aspects of the boundary of the surface water basin or watershed, regardless of the characteristics of streams and reaches within it. Thus, catchment geometry governs surface water flow regime and accumulation. Therefore, the determination of catchment geometry accounts for the geographic dimensions of the highest terrain marks confining the ditched stream networks where from water starts flowing. These dimensions often used to analyze the spatial flow mechanism of the lower-ranking streams to the primary stream, and can thus be used to estimate the intervals for water flow permanence between different streams, as well as to calculate flow volume when a number of related factors is taken into account, specifically the surface slope (Al Saud 2018b).

There are several formula to analyze the catchment geometry, in particular the basin shape is the most significant; and therefore, funnel-like shape basins are the most regular types with uniform water flow, whereas other shape result diverse flow regime.

The following are the fundamental geometric formula used to characterize drainage basins, and applied to NEOM Region.

**Table 5.1** Drainage basins of NEOM Region

| Drainage basin Number[a] | Area ($km^2$) | Perimeter (km) | Principal orientation |
|---|---|---|---|
| **1** | 552 | 144 | NW |
| **2** | 486 | 168 | |
| **3** | 46 | 47 | W |
| **4** | 101 | 86 | |
| 5 | 53 | 48 | |
| 6 | 119 | 92 | |
| 7 | 7434 | 790 | S |
| **8** | 7333 | 830 | SW |
| 9 | 57 | 53 | W |
| 10 | 80 | 69 | |
| 11 | 71 | 51 | |
| 12 | 399 | 147 | SW |
| 13 | 20 | 32 | |
| 14 | 76 | 65 | |
| 15 | 33 | 41 | W |
| 16 | 22 | 33 | |
| 17 | 496 | 112 | SW |
| 18 | 45 | 42 | W |
| 19 | 26 | 35 | NW |
| 20 | 21 | 30 | |
| 21 | 476 | 268 | S |
| 22 | 24 | 33 | NW |
| 23 | 32 | 35 | W |
| 24 | 34 | 41 | SW |
| 25 | 134 | 98 | S |
| 26 | 152 | 92 | SE |
| 27 | 17 | 46 | S |
| 28 | 23 | 38 | |
| 29 | 1198 | 302 | SW |
| 30 | 24 | 35 | S |
| 31 | 54 | 63 | SW |
| *32* | 2382 | 599 | |
| 33 | 47 | 53 | S |
| 34 | 61 | 65 | SW |
| 35 | 23 | 39 | |
| 36 | 198 | 125 | |

(continued)

**Table 5.1** (continued)

| Drainage basin Number[a] | Area (km$^2$) | Perimeter (km) | Principal orientation |
|---|---|---|---|
| 37 | 1464 | 356 | |
| 38 | 98 | 93 | |
| 39 | 131 | 80 | |
| 40 | 331 | 156 | |
| 41 | 164 | 107 | |
| *42* | 672 | 221 | W |
| **43** | 41 | 62 | SW |
| 44 | 17 | 48 | W |
| 45 | 56 | 29 | |

[a]Number according to the produced map (Fig. 5.2)
• Bold numbers: these drainage basins are originated (or partially originated) from NEOM Region and outlet outside it (mentioned area within NEM region)
• Italic numbers: these drainage basins are originated from outside NEOM Region and outlet inside it (mentioned area within NEM region)

### *5.4.1 Catchment Territory*

The territory of the catchment is always given attention as one of the major pillars of the geometric specifications. Thus, territory of a catchment controls water flow regime including mainly water flow velocity and it has a significant role in flood assessment as well as in the erosional processes.

For NEOM Region, the major catchments (i.e. larger than 150 km$^2$) and their territory where streams spread out, were investigated in-depth and thus detailed drainage basin maps were illustrated (Fig. 5.3).

1. Relief gradient ($R_g$):

According to Pike and Wilson (1971), the frequency of dismemberment is expressed by the relief gradient, which is the ratio of upland to lowland elevations within the catchment area. It is also described as the land mass maturity, and thus youth stage basins show intervals of dissection skewed toward the lower altitudes and vice versa.

$R_g$ is calculated using the mean elevation ($E_{mean}$), maximum elevation ($E_{max}$) and minimum elevation ($E_{min}$), according to the following formulae:

$$R_g = \frac{E_{mean} - E_{min}}{E_{max} - E_{min}}$$

Therefore, For NEOM Region, it was found that the average relief gradient is about 0.5 which is almost a moderate gradient (Table 5.2).

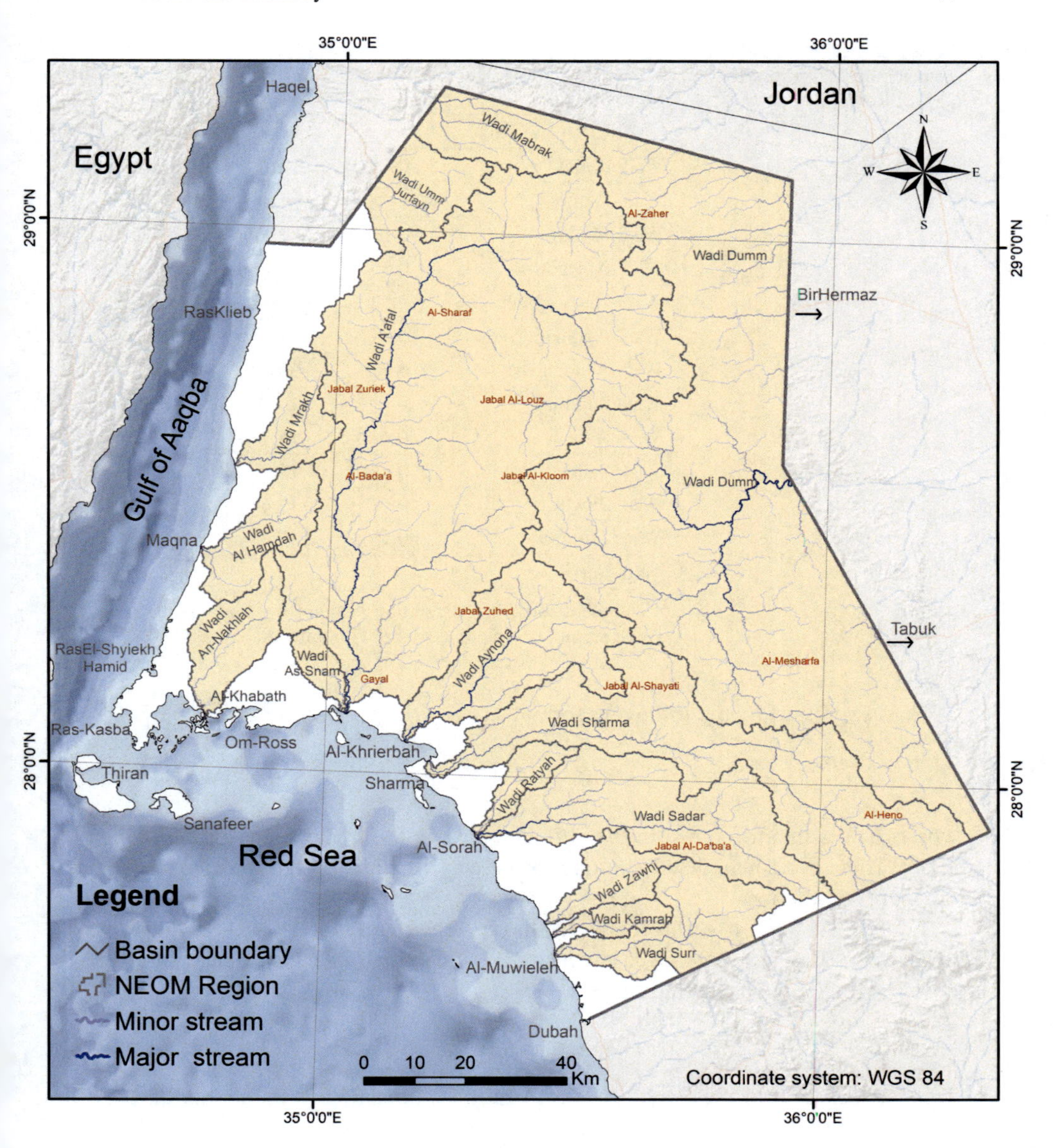


**Fig. 5.3** Drainage basins (catchments and streams) of NEOM Region

2. Relief ratio ($R_r$):

This geometric specification represents the average slope in the entire basin, which is almost a function of topographic gradient of the basin territory. According to Chorley et al. (1984), $R_r$ controls infiltration, moisture in near-surface saturated zones, as well as it enhances the lateral flow of groundwater.

Table 5.2 Major formula of the catchments territory for NEOM Region

| Drainage basin | $R_g$ | $R_r$ | $C_s$ |
|---|---|---|---|
| Wadi Mabrak[a] | – | – | – |
| Wadi Umm Jurfayn[a] | – | – | – |
| Wadi A'afal | 0.41 | 20.7 | 1.3 |
| Wadi Dumm[a] | – | – | – |
| Wadi Mrakh | 0.72 | 42.3 | 5.6 |
| Wadi Al Hamdah | 0.56 | 30.9 | 1.1 |
| Wadi An-Nakhlah | 0.52 | 6.12 | 1.8 |
| Wadi As-Snam | 0.39 | 9.14 | 1.5 |
| Wadi Aynona | 0.38 | 30.4 | 1.1 |
| Wadi Sharma[a] | – | – | – |
| Wadi Ratyah | 0.46 | 15.7 | 1.3 |
| Wadi Sadar | 0.47 | 18.2 | 1.6 |
| Wadi Zawhi | 0.63 | 17.6 | 1.5 |
| Wadi Kamrah | 0.55 | 19.4 | 1.4 |
| Wadi Surr[a] | – | – | – |

[a]Catchment territory calculations cannot be applied for basins which are not completely extend within the area of study

$R_r$ is expressed by the following formulae:

$$R_r = \Delta\text{H}/\text{L}_\text{b}$$

where $\Delta H$ is the difference in the altitude between the highest and lowest points, and $\text{L}_\text{b}$ is the horizontal distance along the longest dimension of the basin parallel to the main stream line.

For NEOM Region, it was found that the average relief ratio is 21.04, while there is obvious difference that ranges between 6.12 and 42.3 (Table 5.2). Hence, all relief ratio above 20 are considered as high.

3. Mean catchment slope ($C_s$):

This is measured for the catchment upstream of each sampling site by dividing the difference in elevation between specific points over length of the catchment. It is a function of high rate of the overland flow and vice versa.

Morisawa (1976) used the following formulae to calculate $C_s$:

$$C_s = \frac{(\text{E } 0.85 \text{ L}) - (\text{E } 0.10 \text{ L})}{\text{E } 0.75 \text{ L}}$$

where E is the elevation, L is the catchment maximum length, and points are taken along this line (0.10 L near the lower part of the catchment, 0.85 L towards the upper end). Hence, the slope (in degree) = $\tan^{-1}$ (slope in decimal form).

For NEOM Region, all catchment slopes are moderate, and averaging at about 1.82 except that for Wadi Mrakh which is 5.6 (Table 5.2).

### *5.4.2 Catchment Shape*

The catchment shape, as an orientation for the outer boundary of a drainage basin, has an essential role on several hydrological processes occur in the basin. Thus, there are many shapes of catchments that are supposedly reflecting run-off "bunch up" at the outlet (Al Saud 2009). For example, the ultimate concentration of its peak flow can be used to help in studying the effects of catchment shape on the hydrograph and on stream behavior (Black 1991). Thus, a catchment with circular shape, for, example, would result in run-off from various tributaries of the drainage basin reaching the outlet at the same time (Fig. 5.4). While, an elliptical catchment having the outlet at one end of the major axis and having the same area as the circular watershed would cause the runoff to be spread out over time, thus producing a smaller flood peak than that of the circular watershed (Das 2000).

1. Basin maximum length ($L_{max}$):

This is the length between the two extreme points on the basin along the primary stream. $L_{max}$ is a function of the topographic framework of the basin and directly affects the time lag of water flow from the highest points on the basin to the outlet. Therefore, $L_{max}$ work in accelerating water flow energy and then controls time of leakage, evaporation, and transpiration (Al Saud 2018a, b). It geometric specification

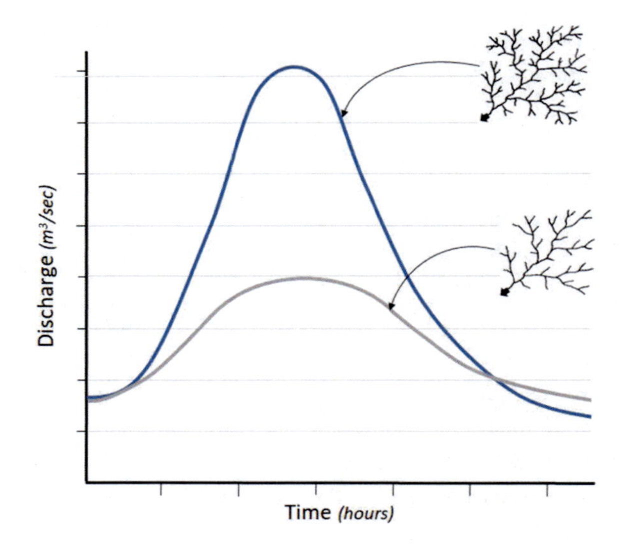


**Fig. 5.4** Example showing the basin effects on hydrograph shape

is calculated as a straight line along the main course from the highest point to the outlet. Table 5.2 shows the maximum length in basins on NEOM Region.

It is obvious that $L_{max}$ in NEOM Region ranges between 5 and 10 km, with an average $L_{max}$ of approximately 7 km. This indicates that For NEOM Region, it was found that the average basin length is 35.7 which are relatively small to moderate length, but some show considerable length like Wadi A'afal and Wadi Sadar (Table 5.3).

2. Length/Width rations ($Lw_r$):

This ration between the lengths (L) of a basin to its width (W), and it is often calculated because it evidences the water flow energy between upstream and downstream. Thus, the length/width ration is a function of time that run-off effectively reached the primary watercourse (Al Saud 2009). In the typical basins, the average $Lw_r$ ration is 0.5, which indicates that the basin length is almost twice equal its width (i.e. approximately funnel-like shape). Also, the higher the $Lw_r$ ration, the higher run-off duration is and sufficient time lag for water infiltration to the substratum.

Hence, $Lw_r$ can be calculated as follows:

$$Lw_r = \mathrm{L/W}$$

For the basins in NEOM Region, $Lw_r$ were calculated as shown Table 5.3. It is obvious that the average Length/Width ration in NEOM Region is 3.89 which indicates that most catchments have normal orientation as a funnel-like shape.

**Table 5.3** Major formula of the catchments shape for NEOM Region

| Drainage basin | $L_{max}$ | $Lw_r$ | $E_i$ | $F_f$ | $G_i$ | $C_r$ |
|---|---|---|---|---|---|---|
| Wadi Mabrak | – | – | – | – | – | – |
| Wadi Umm Jurfayn | – | – | – | – | – | – |
| Wadi A'afal | 123.4 | 2.86 | 0.79 | 0.49 | 5.16 | 0.047 |
| Wadi Dumm | – | – | – | – | – | – |
| Wadi Mrakh | 26.1 | 3.24 | 0.86 | 0.58 | 4.15 | 0.074 |
| Wadi Al Hamdah | 32.6 | 4.01 | 0.77 | 0.47 | 2.84 | 0.158 |
| Wadi An-Nakhlah | 37.7 | 2.93 | 0.65 | 0.33 | 6.93 | 0.076 |
| Wadi As-Snam | 24.2 | 3.44 | 0.57 | 0.26 | 8.94 | 0.072 |
| Wadi Aynona | 52.9 | 3.52 | 0.74 | 0.43 | 4.92 | 0.052 |
| Wadi Sharma | – | – | – | – | – | – |
| Wadi Ratyah | 24.6 | 4.93 | 0.64 | 0.39 | 4.94 | 0.050 |
| Wadi Sadar | 71.1 | 4.37 | 0.61 | 0.29 | 5.25 | 0.046 |
| Wadi Zawhi | 26.2 | 3.31 | 0.78 | 0.48 | 4.84 | 0.054 |
| Wadi Kamrah | 25.8 | 6.34 | 0.56 | 0.25 | 4.71 | 0.057 |
| Wadi Surr | – | – | – | – | – | – |

3. Elongation Index ($E_i$):

It describes the ratio between the diameter of the circle with the same area as the catchment of the drainage basin and the distance between the most extreme two points in the catchment. The following formulae describes $E_i$:

$$E_i = \frac{2\sqrt{\mathrm{A}}}{\mathrm{L_m}\sqrt{\pi}}$$

According to Schumm (1956), $E_i$ was classified as: <0.5, 0.5–0.7, 0.7–0.8, 0.8–0.9 and 0.9–1 for more elongated, elongated, less elongated, oval and circular; respectively.

For NEOM Region, it was found that the average elongation index is about 0.69, and this is nearly consistent with the results of $Lw_r$ (Table 5.3).

4. Form Factor ($F_f$):

It expresses the ratio of the basin to its axial length where the latter is the horizontal distance along the maximum basin dimension parallel to the main stream Horton (1932). It is; therefore, a function the shape of the catchment, its profile and channel dimension. $F_f$ indicates the flow intensity within the catchment.

According to Horton (1932), $F_f$ less than 0.19 indicates elongated catchment shape, while $F_f$ greater than 0.53 indicates a circular shape. Moreover, higher $F_f$ value experience larger peak flows with shorter duration.

Therefore, $F_f$ is represented by the catchment area (A) and length (L) as follows:

$$F_f = \frac{\mathrm{A}}{\mathrm{L}^2}$$

For NEOM Region, the average r form factor is 0.39 and all resulted calculations indicated almost normal catchment forms (Table 5.3).

5. Gravelius index ($G_i$):

Gravelius (1914) defines $G_i$ as the ration between the outer catchment boundary (i.e. perimeter) and circle with the same area as that of the catchment. Thus, $G_i$ is always greater than 1. However, $G_i$ close to 1 indicates an elongated catchment.

$$G_i = P/\sqrt{\pi A}$$

For NEOM Region, Gravelius index average sat about 5.27 which evidences a semi-circle catchment orientations for all catchments. However, some catchments (Wadi An-Nakhlah and Wadi As-Snam) are almost with semi oval orientation (Table 5.3).

6. Circularity ratio ($C_r$):

Miller (1953) calculated $C_r$ as the degree of similarity between the outer limits of the catchment with respect to a circle. Thus, it compares between the area of the catchment to the area of a circle have the same perimeter. It is calculated as:

$$C_r = 4A/P^2$$

Hence, when the value of $C_r$ is exactly 1 the catchment is set to be a perfectly circular shape, $C_r$ with less than 1 indicates that the catchment much more elongated. Miller (1953) described the basin of the circularity ratios range 0.4–0.5 as a strongly elongated and highly permeable homogenous geologic materials. Also, the elongated catchments often have low discharge of runoff and highly permeability of the subsoil condition. While, Waikar and Nilawar (2014) attributed high value of catchment circularity ratio to the late maturity stage of topography.

For NEOM Region, the circularity ratio average at about 0.6, and it is in consistency with Gravelius index (Table 5.3).

## 5.5 Stream Morphometry

The morphometry of the drainage basin, in general and stream network in particular, belongs to the arrangement, interrelationship and orientation of streams, including the primary and secondary tributaries and even the existing reaches. This hydrologic specification is controlled mainly by the geology and geomorphology within the catchment as well as the dominant climatic conditions. Thus, the morphometry of drainage basin can be considered as more significant than its geometry. Therefore, many formula are applied to evaluate the stream morphometry, which are always used while surface studying water flow and storage regime.

1. Stream density ($D_d$):

The most important in drainage distributions is the density of tributaries among the catchment (Al Saud 2012). Thus, the density of streams has been well identified in several geomorphological studies especially that this hydrologic phenomenon is a function of infiltration/run-off rate. Therefore, dense streams evidences low infiltration rate, which results in surface run-off and overland flow. Hence, the calculation of $D_d$ is a useful numerical measure of landscape analysis and run-off potential (Chorley 1969).

According to Horton (1945), stream density is a measure of the total length of stream channels ($L_S$) per unit area of drainage basin. Mathematically it is expressed as:

$$D_d = \sum \mathrm{L_s}/\mathrm{A}$$

To create drainage density map, drainage basins were classified into frame with defined area, and the total length of streams in each area was calculated. Thus, the resulted values were plotted on a map to construct a contour maps for the different stream densities.

Therefore, each drainage basin can be classified, according to it stream density, as follows:

- Coarse drainage basin <5 km/km$^2$
- Medium drainage basin 5–10 km/km$^2$
- Fine drainage basin 10–20 km/km$^2$
- Very fine drainage basin >20 km/km$^2$.

For NEOM Region, it can be considered that all drainage networks are of the coarse type since all of them are below 5 km/km$^2$ (Table 5.4).

2. Meandering ration ($M_r$):

A stream often have several meanders while it spans on terrain surface, and therefore, different aspects and dimensions of meanders appears within the same water channel. This depends on several geomorphologic and hydrological characteristics such as slope gradient, rock types, geological formations and many others (Al Saud 2012).

**Table 5.4** Morphometric measurements of drainage basins in NEOM Region

| Drainage basin | $D_d$ (km/km$^2$) | $M_r$ | $T_t$ (streams/km) | $S_c$ (m/km) |
|---|---|---|---|---|
| Wadi Mabrak[a] | 1.86 | 0.87 | 7.13 | – |
| Wadi Umm Jurfayn[a] | 1.75 | 0.94 | 5.06 | – |
| Wadi A'afal | 1.54 | 2.07 | 14.49 | 10.45 |
| Wadi Dumm[a] | 1.39 | 2.42 | 12.28 | – |
| Wadi Mrakh | 0.56 | 0.54 | 1.52 | 32.06 |
| Wadi Al Hamdah | 1.07 | 0.58 | 4.48 | 14.06 |
| Wadi An-Nakhlah | 1.02 | 0.45 | 1.81 | 9.28 |
| Wadi As-Snam | 0.26 | 0.31 | 0.42 | 3.75 |
| Wadi Aynona | 0.49 | 0.68 | 1.94 | 22.9 |
| Wadi Sharma[a] | 1.64 | 0.89 | 6.52 | – |
| Wadi Ratyah | 1.51 | 1.01 | 2.39 | 9.76 |
| Wadi Sadar | 1.67 | 0.97 | 6.86 | 3.52 |
| Wadi Zawhi | 0.37 | 0.34 | 0.78 | 9.75 |
| Wadi Kamrah | 0.48 | 0.42 | 0.73 | 10.32 |
| Wadi Surr[a] | 0.28 | 0.47 | 0.85 | – |

[a]The stream slope could not be calculated because the total length of these e basins are not included in NEOM Region

Usually, the meandering ration plays a role in flood assessment, since the increase in the meandering ratio decrease the flow energy and increase the stream load capacity due to erosion mechanism on the meandering sites (Al Saud 2012).

Meandering ration is calculated by comparing the ratio between the straight length ($L_s$) of the main stream and its length within the existing curvatures ($L_m$) according to the following equation:

$$M_r = L_s/L_m$$

For NEOM Region, the average meandering ratio is 0.86 which evidences approximately ordinary meandering values, but it is high in Wadi A'afal and Wadi Dumm (Table 5.4).

3. Texture topography ($T_t$):

This formulae expresses the degree of drainage dissection. Thus, it indicates the ability of terrain to shrink water, which is influenced by lithology and structure (Al Saud 2009). It is also expressed as the spacing between streams, and it is used in determining the quantity of relief which vary greatly with differences in stream spacing (Douglas 1933).

Accordingly, Smith (1950) classified texture topography into three main categories. They are: soft, moderate and rough for $T_t < 4$, 4–10 and >10; respectively.

The variables used to calculate are the number of streams in a catchment ($S_n$) and the perimeter of this catchment (P). Therefore, the following equation defines the texture topography:

$$T_t = \sum S_n/P$$

For NEOM Region, the texture topography average about 4.48 streams/km which accords with meandering ratio, notably that higher $T_t$ values are in Wadi A'afal and Wadi Dumm (Table 5.4).

4. Mean stream slope ($S_c$):

This simplified morphometric specification estimates the slopping of the primary stream in a catchment. It interrelates with many of the previous specifications where several geologic and geomorphologic parameters play a role in the slopping rate of the primary stream. Therefore, $S_c$ is calculated by dividing the difference in altitude between the source point (i.e. most upstream elevated reaches) and the outlet divided by the total stream length. Therefore, the higher $S_c$ enhances the flow (i.e. run-off) rate along the primary stream and vice versa. Hence, $S_c$ is represented by the following formulae according to Raven et al. (2000):

$$S_c = E_s - E_o/L$$

where $E_s$ is the elevation at sources and $E_o$ is the elevation at the outlet.

For NEOM Region, the average value of the mean stream slope is 12.58 m/km, which indicates a gentle to moderate slope of the primary watercourses, except for Wadi Mrakh which is high with 32.06 m/km Table 5.4.

## References


Aawari, I. (2005). Vegetation cover in Wadi Noaman with special emphasis on its tributary: Wadi El Majayrish. Unpublished MSc thesis (in Arabic). Department of Geography, Faculty of Arts. Jiddah, KSA, pp. 473.

Abdelkarim, A., Gaber, A., Youssef, A., & Prandhan, B. (2019). Flood hazard assessment of the urban area of Tabuk city, kingdom of Saudi Arabia by integrating spatial-based hydrologic and hydrodynamic modeling. *Sensors (Basel), 19*(5), 1024. https://doi.org/10.3390/s19051024.

Al-Momani, A., & Shawaqfah, M. (2013). Assessment and management of flood risks at the city of Tabuk, Saudi Arabia. *The Holistic Approach to Environment, 3*(1), 15–31.

Al Saud, M. (2009). Watershed characterization of Wadi Aurnah, Western Arabian Peninsula. *Journal of Water Research and Protection (JWARP), 1*, 316–324.

Al Saud, M. (2011). The role of space technology and geo-informatics in the water strategies: applications from Saudi Arabia Scientific Forum on: Arab Water Security Strategy, which was organized by the Faculty of Science strategy—Prince Naif Arab University for Security Sciences, December 19–21.

Al Saud, M. (2012). Use of Remote Sensing and GIS to Analyze Drainage Basin in Flood Occurrence, Jeddah-Western Saudi Coast. Book on "Drainage basins". InTech.

Al Saud, M. (2018a). Using Space Techniques and GIS to Identify Vulnerable Areas to Natural Hazards along the Jeddah-Rabigh Region, Saudi Arabia. Nova Science Publisher Inc., New York, p. 306. ISBN13: 978-15-361-33134.

Al Saud, M. (2018b). *Geomorphological Characteristics of Drainage Basins in the Riyadh Region A Focus on Wadi Al-Saly Basin* (p. 138p). Massachusetts: Publishing Solutions Group.

Al-Saud, M. (2007). Using satellite imageries to study drainage pattern anomalies in Saudi Arabia. *Journal of Environmental Hydrology, 15*(30), 1–15.

Black, P. (1991). *Watershed Hydrology* (p. 324). NJ: Prentice Hall Advanced Reference Series.

Chorley, R. J. (1969). Introduction to physical hydrology. Methuen and Co. Ltd., Suffolk, p. 211.

Chorley, R., Schumm, S., & Sugden, D. (1984). *Geomorphology* (p. 607). London: Methuen.

Das, G. (2000). *Hydrology and Soil Conservation Engineering*. New Delhi, India: Prentice Hall of India.

Douglas, J. (1933). Available relief and texture of topography a discussion. *The Journal of Geology. The University of Chicago Press, 41*(3), 293–305.

Gravelius, H. (1914). Rivers. G.J. göschen Publishing. Berlin, p. 179.

Horton, R. E. (1932). Drainage-basin characteristics. *Transactions American Geophysical Union, 13*, 350–361.

Horton, R. (1945). Eroasinal development of streams and their drainage basins; hydrophyscal approach to quantitative morphology. *Geological Society of America Bulletin, 56*, 275–370.

Miller, V, (1953). A quantitative geomorphic study for drainage basin characteristics in the Clinch Mountain area, Virginia and Tennessee, Technical Report No. 3, Geology Depart., Colombia University, I-30, N6 ONR 271–30.

Monteiro, E., Fonte, C., & Lima, J. (2018). Analysing the potential of openstreet map data to improve the accuracy of SRTM 30 DEM on derived basin delineation, slope and drainage network. *MDPI Publishing. Hydrology, 5*, 34.

Morisawa, M. (1976). Geomorphology Laboratory Manual. Wiley Inc. N.Y., pp. 1–253.

Pike, R., & Wilson, S. (1971). Elevation-relief ratio. Hypsometric integral and geomorphic area-altitude analysis. *Geological Society of America Bulletin, 82*, 1079–1084.

Qari, M. (2009). Geomorphology of Jeddah Governate, with emphasis on drainage basins. *Journal of King AbdulAziz University (JKAU): Earth Science, 20*(1), 93–116.

Raven, P., Holmes, N., Naura, M., Dawson, F. (2000). Using river habitat survey for environmental assessment and catchment planning in the U.K. *Hydrobioloigia, 422*(0), 359–367.

Schumm, S. (1956). The elevation of drainage basins and slopes in badlands at Perth Amboy, New Jersey. *The Geological Society of America Bulleti, 67*, 597–646.

Smith, K. (1950). Standards for grading texture of erosional topography. *American Journal of Science, 248*, 655–668.

Subyani, A., Qari, M., Matsah, M., Al-Modayan, A., & Al-Ahmadi, F. (2009). *Utilizing remote sensing and GIS technologies to produce hydrological and environmental hazards in some Wadis, western Saudi Arabia (Jeddah-Yanbua)*. KSA: Department of Hydrology. King Abdulaziz City for Science and Technology.

Waikar, M., Nilawar, A. (2014). Morphometric analysis of a drainage basin using geographical information basin: A case study. *International Journal of Multidisciplinary and Current Research, 2* (Jan/Feb 2014 issue).

# Chapter 6
# Potential Natural Resources

**Abstract** Many countries worldwide have become with an outstanding economical rank and at the best level of development due to their natural resources. Hence, these resources, whether renewable or non-renewable, comprise the majority of the real wealth of nations. Natural resources contribute towards the increasing the capital and income, as well as they stabilize fiscal revenue and act on poverty reduction. However, natural resources are not always viewed from the economic aspect of view, but in many instances they are necessary to build developed regions. In other words, the presence of natural resources contribute in creating several development activities, especially in the establishment of smart zones with economic potentials. This is the case for NEOM Region which has already started building the bases to be the global hub for the entire Middle East Region, and this has been dependent largely on its available natural components, including the natural resources. These resources are still considered as neglected treasures so far. Therefore, identifying the available natural resources is essential for optimal SLM, because it will provide pillars for several aspects of development in NEOM Region. This implies job opportunities, availability of vital resources such as water and crops, touristic sites, as well as accelerating the economical cycle in the region. This chapter presents the importance of identifying the available natural resources with a special emphasis on groundwater and ore deposits as the major two resources in this region.

**Keywords** Groundwater · Geologic assessment · Employment · GDP · Ore deposits · Economic zone

## 6.1 Introduction

Usually there is contradictory about the definition of natural resources, and the common understanding is that natural resources are derived from beneath Earth's surface, like petroleum and ore deposits. Nevertheless, natural resources have a variety of materials including natural elements that exist without actions of humankind to their origin, and they used to support life and meet of people needs where some of them are vital for life such as water and plants.


M. M. Al Saud, *Sustainable Land Management for NEOM Region*,
https://doi.org/10.1007/978-3-030-57631-8_6

Up to date, there are no clear guidelines for classifying natural resources to assess precisely how the economic incentives and opportunities differ for various natural resources. Therefore, the insufficient classification of resource types and lack of data has impeded statistical research and the identification of resources type, location and duration (Lujala 2003).

Most natural resources are of the organic sources and elaborated by human to be usable. Therefore, they include mainly: water, air, sunlight, soil, plants, petroleum, coal, metallic and non-metallic ores and they may extend to many other resources on Earth's surface. Hence, some of these resources are renewable and others and not.

Using natural resources to promote economic development sounds in continues advance, whereas, countries which have subsoil assets such as hydrocarbons and minerals, are seeking to transform them into surface assets and invest them in human and physical capital which can be used to support employment and increase economic growth. Indeed, this transformation has proved to be hard practically (Venables 2016). Thus, few developing economies succeeded with this transformation, and economic growth has generally been lower in resource-rich developing countries than in those without enough resources. It was not until the 2000s (i.e. the period of rising commodity prices) that resource-rich countries grew faster, although even then per capita growth was the same in both categories of countries (Venables 2016). Lately, this concept has been oppositely acted, and it has been evidenced by the establishment of smart cities in many resource-rich developing countries, such as the case in Singapore, Jakarta in Indonesia and New Delhi in India.

Looking to the list of top 50 smart cities worldwide; however, not all of them are located in the proximity of zones with available natural resources, but some of them are so. In fact, the presence of natural resources in the surrounding of these cities added significant value to their ranking and used in the creation of the Smart Evaluation Method (SEM) which is often developed aiming at responding to the create sustainable environment and reduce negative consequences and the existed challenges.

The existence natural resources (e.g. mineral resources, environmental resources, etc.) in any region enable opening new opportunities for labor, accelerating the economic cycles as well as helps in making this region as a hub for trading and commercial exchanges.

### *6.1.1 Natural Resources in NEOM Region*

Similar to most of states of the Gulf Cooperation Council (GCC), attention for natural resources in the Kingdom of Saudi Arabia is given to petroleum (oil and gas), whereas other resources are available abut with less investment if compared with petroleum. Hence, the availability of petroleum is the major reason why these countries have witnessed a quantum leap, since early 1960s, in the economic and development fields where the Kingdom of Saudi Arabic occupies the outstanding position.

Overall, the Kingdom of Saudi Arabia, the 3rd richest in natural resources, has about \$34.4 $\times$ $10^6$ million worth of natural resources. Other than petroleum, Saudi Arabia's natural resources include copper, feldspar, phosphate, silver, sulphur, tungsten, and zinc (Ivestopedia 2020).

The general figure and understanding to natural resources in the Saudi Arabia imply the areas where oil fields are located in the eastern part of the Kingdom. This can be also the case for some mines of ore deposits, notably those for gold, silver and copper ores which are scattered in different regions of the Kingdom and mainly in the rocks of the Arabian Shield. Nevertheless, no obvious highlights can be noticed for the upper north-western part of Saudi Arabia where NEOM Region is situated, and no studies comprehensive and creditable studies have been performed for this region yet.

No doubt that NEOM Region encompasses several renewable and non-renewable natural resources and this is preened by its unique landscape and purity of its nature. Therefore natural resources in NEOM Region can be allocated under the following categories:

1. Biodiversity

This category of natural resources generally implies the biological features, and more certainly the existing fauna and flora whether terrestrial or marine environments. In fact studies on these features have not investigated in details yet and few number of reconnaissance have been applied. (Examples: Almutairi et al 2016; Aloufi and Amr 2019).

2. Natural exposures

Natural exposures in this section include mainly air and sunlight which can be found everywhere, but their exposure in NEOM Region is distinguished because air is totally pure, and blowing either in the vast open area or in valleys corridors. While sunlight exposure is almost with typical aspects, notably that the region is almost an open space. However, no studies have been done on these exposures, except some meteorological measurements from the available stations (Examples: GAMEP 2019; Meteoblue 2020).

3. Groundwater

In NEOM Region, more than 45% of the exposed lithologies are composed of sandstone rocks and the related clastic facies which are characterized by potentiality for groundwater storage. They are attributed to several overlaying rock formations with a total thickness exceeding 3000 m in total. In addition, the alluvial deposits along the flood plain of wadis are potential lithologies for groundwater storage.

Even though, rainfall rate is little enough (<60 mm/year), but water from rainfall directly infiltrates into these sandstone rocks with very minimal rate of run-off or evaporation.

4. Ore deposits

NEOM Region is categorized as one of the 18 recognized mineralized districts in the Kingdome according to Nehlig et al. (1999). The region is characterized by dominance of metamorphosed and plutonic rock derivatives and complexes, which are almost belong to many intrusions of the Paleozoic Era, are rich with metallic and non-metallic ores. Most of these ores are feasible for quarrying due to the structural complexity where several exposure occur.

According to Clark (1987), the jaspilitic iron ore related to the Wadi Sawawin deposits, the mineralized Jabal At-Tawileh microgranite, and the large quantities of gypsum-anhydrite of the Maqna massif have the greatest economic potential in the region.

5. Potential hydrocarbon resources

Oil and gas are potential hydrocarbon resources in NEOM Region. This concept is based on the: (1) lithological characteristics of the region including mainly the dominant porous and permeable rocks, i.e. sandstone sequences (2) the presence of cap rocks which are represented by the existing evaporates (e.g. gypsum and anhydrite) and (3) existence of intensive structures (e.g. folding and faulting).

In this respect, an aeromagnetic survey has been carried out in 1962 in Lisan basin underlying Wadi A'afal and extending off-shore (Agocs and Keller 1962). The Tertiary and Quaternary deposits in this plain exceed 3000 m, as well as oil seeps are known along the coastline between Al Wajeh and Ras El-Shyiekh Hamid. Therefore, 7 test wells were drilled off-shore in the area between Al-Khrierbah-Al Bada'a and Al-Muieleh (Ahmad 1972; Coleman 1977).only one well showed positive results and penetrated through three sandstone reservoirs, one of which at 2600 m yielded good quality gas and 600 barrel per day of condensate.

## 6.1.2 *Management of Natural Resources*

Natural resources sustain a complicated interaction between living things and non-living things where human is the first beneficiary from this interaction. Thus, more than 60 billion tons of resources world are used by economy each year to produce the goods and services. On the average, a person consumes about 35 kg of resources per day (SERI and GLOBAL 2009).

The optimal management of natural resources remains challenging, especially in the time the economical oscillations and competitions are tremendous due to many factors including mainly the geopolitical conflicts, shared resources, mismanagement and the lack to adequate financial resources. Therefore, successful management needs planning and preparation for all incentives and tools required. This often starts from the beginning when resources are identified till the final benefit. In this respect, there are many stages applied for best management as follows:

1. Resources recognition

Usually, the recognition of resources is the initial step for further implementations. Thus, the identification can happened spontaneously (by chance) or by noticing remarkable features evidence the presence of resources, or by applying investigations and technical surveys.

2. Dimensional aspects of resources

The identified natural resources need more investigation in order to assess their dimensional extent (e.g. reserve, sustainability, etc.) which is a function of investment feasibility. This is significant to figure out whether these resources can be invested. For example, limited petroleum resources will not considered for giant projects (e.g. construction of refineries, etc.) related to oil and gas.

3. Feasibility and economic valuation

Feasibility assessment is very important step to begin in the investment of natural resources. This will include several components, in addition to the above mentioned dimensional aspects and resources reserve and sustainability, the accessibility and workability of using resources. For example, sometimes ore deposits are available but it is not accessible to be reached, or if the impurities in such ores are abundant, this also makes a constraints for investment. Also, if the cost of quarrying and production of ores exceed the economic value of resources, thus it will not a feasible resources for exploitation.

4. International marketing and economic mobilization

One of the major factors of successful economy is the dependence of local production, and this can be done by the increasing investment and the mobilization of financial resources. Therefore, the economic growth is controlled by the capital market and the mobilization of capital formation. This is an economical process where the involvement of the international marketing will play a significant role.

5. Urban mobilization

Benefit realization is greater in urban areas, where projects have been based on co-creation and urban mobilizations. Therefore, new approaches for urban projects are essential in order to reach strong visions and the best intentions of urban projects. Thus urban mobilization can be reached by the management and involvement of stakeholders in the right processes with the right timing, as well as the actual co-creation with citizens and the communication during the early development phases.

Therefore, the optimal management of natural resources, in a Region liked NEOM, can provide and contribute in the following:

- Rise the standard of living and attaining prosperity and acting on poverty reduction.
- Creation of employment and job potentials.
- Optimal investment of ecosystem and natural features.
- Create a hub for international cooperation and world peace and motivating cultural exchange.
- Ensures optimal and rational and wise-use of resources
- Increasing and accelerating the industrial growth and create compete trading.
- Better managing of surplus production through the involvement of international marketing.
- Motivating foreign investments and mobilization of resources.

For NEOM Region, there are many aspects of natural resources to be exploited. Two of them are directly related to the geology of the region and they will be discussed in this chapter as significant geological natural resources. Whereas, other natural resources are related to the ecosystems which has an important impact on the natural resources (will be discussed in next chapter) and land management. Thus, the geological ones are: (1) groundwater and (2) ore deposits.

## 6.2 Groundwater Resources

Groundwater is water resources that are found in rocks, as aquifers, at different depths below the surface of the Earth, and it is usually considered as one of the most important natural resources. This aspect of water resources has the advantage of good quality and it remains mostly in stored static condition. Besides, the exploration of groundwater is still challenging since it is an invisible source.

Groundwater in NEOM Region is utmost significant, notably that this region receives very limited rainfall (i.e. <60 m/year). Therefore, the flowing surface water is rapidly infiltrates into substratum in this region, because all the biggest part of the exposed rocks are characterized by potential hydraulic properties towards high porosity and permeability, both as primary and secondary hydraulic properties.

No studies have been applied on groundwater resources for the geographic part of the most northwest of Saudis Arabia where recently NEOM Region has been designated. However, there are some studies obtained but on Tabuk Region which is hydrogeological similar to the plateau of NEOM Region, but totally different from the other parts of this region (Al-Ahmadi 2009; UN-ESCWA and BGR 2013).

Only a briefing on the hydrogeology of Al-Bada'a quadrangle was mentioned by Clark (1987). Thus, it was mentioned that groundwater is obtained from shallow wells in several major wadis and is used locally for irrigation purposes and for livestock, in general it is scarce. however, significant quantitates of groundwater are available in the plateau and extending near Tabuk where various aquifers are found in the Paleozoic sandstone, in addition, similar aquifer may exist where the Paleozoic rocks are exposed in within NEOM Region.

### *6.2.1 Hydrogeology of NEOM Region*

The sharp diversity in rock lithologies as well as the complicated geology of NEOM Region characterizes its hydrogeological features and then controls groundwater flow and storage regime. Thus, NEOM Region occupies different rock types which are characterized by different rock facies and then diverse hydraulic properties. This in turn resulted different hydrogeological features between the rock sequences in the region. Hence, there are diverse aquiferous attributes for each of these formations. These aquiferous attributes describe the controlling hydrogeological characteristics of rock sequence in the region, and they can be under the flowing attributes:

- Aquifer: Rock formation, rock unit or group of formations that store groundwater with considerable amounts and can be tapped successfully.
- Aquiclude: Rock formation or rock unit which is porous and capable of storing water, but it does not transmit it at rates sufficient to furnish an appreciable supply for a well or spring (WMO 1974).
- Aquitard: A confining bed that retards but does not prevent the flow of water to or from an adjacent aquifer. It does not readily yield water to wells or springs, but it can be a groundwater storage unit (AGI 1980).
- Aquifuge: A rock unit which contains no interconnected openings or interstices, and therefore neither stores nor transmits water (ASCE 1985).

Even though, the plotted rock formations and units were calculated at 80 ones, as they were mentioned in Sect. 3.4.1, yet the integration and merging of different units and formations resulted only 53 rock formations as they are plotted in Table 6.1.

Based on the above aquiferous attributes; therefore, the geological rock succession of NEOM Region can be illustrated as shown in Table 6.1. The obtained illustration is based mainly on the lithological characteristics of the existed rocks which are from the Pre-Cambrian to Quaternary in age. In particular, the illustration of the aquiferous attributes accounted the following hydrogeological elements:

1. The presence of primary and secondary porosity and permeability, as major hydraulic properties.
2. Lithological characteristics, with a special emphasis to rock type including hardness and compaction.
3. The contents of argillaceous materials which are major component in creating porous but not permeable rocks.
4. The dominant rock deformations, and certainly the fracture systems including fissures and joints.
5. Presence of rock dissolution processes, such as karstification where conduits and cavities exist in the carbonate rocks.

It is obvious from the illustrated Table 6.1 that the largest part of the rock succession in NEOM Region is either aquifer or Aquifuge rock formations. Thus, the majority of aquifer rocks begins from the Cambrian period where sedimentary environment of deposition took place with Siq and Quweira and Ram and Umm

**Table 6.1** Aquiferous attributes of rocks in NEOM Region

| Period | Rock formation/lithology | Thickness[a] (m) | Aquiferous attributes |
|---|---|---|---|
| Quaternary | Undifferentiated sand and gravel, Aeolian sand dunes and sand sheets, gravel and terraces, talus and alluvial deposits and Sabkha | Variable and almost less than 100 m | Locally aquiferous |
| Tertiary and quaternary | Conglomerates | | |
| | Lisan formation | 3000 | Aquifer |
| Tertiary | Bad formation | 300 | Aquitard |
| | Nutaysh formation | 400 | Aquifer and partially Aquitard |
| | Musayr formation | 120–150 | |
| | Usayliyah formation | 25–50 | Aquifer |
| | Khuraybah formation | 250–300 | Aquiclude |
| | Sharik formation | 1000 | Aquifer |
| | Azlam formation | 250 | Aquiclude |
| | Adaffa formation | 200 | Aquifer and partially Aquiclude |
| Cretaceous and tertiary | Gabbro and diorite dikes | Undefined | Aquifuge |
| Ordovician-Silurian and Devonian | Tabuk formation | 1000–1100 | Aquifer and partially aquiclude |
| Cambrian - Ordovician | Ram and Umm Sahm sandstones | 400 | |
| Cambrian | Quweira sandstone | 200 | Aquifer |
| | Siq sandstone | 25–50 | |
| | Sawda complex | Undefined | Aquifuge |
| Pre-Cambrian (Overlapped rocks with unconformities) | Undifferentiated volcanics | | |
| | Lawz complex | | |
| | Minaweh formation | 500 | Aquiclude |
| | Sulaysiyah formation | Undefined | Aquitard |
| | Saluwah formation | | Aquitard |
| | Shar complex | | Aquifuge |
| | Dabbagh complex | | |
| | Massah complex | | |
| | Mabrak granite | | |
| | Maharish complex | | |

(continued)

**Table 6.1** (continued)

| Period | Rock formation/lithology | Thickness[a] (m) | Aquiferous attributes |
|---|---|---|---|
| | Sadr complex | | |
| | Sawawin complex | | |
| | Duba complex | | |
| | Midyan suite | | |
| | Mowasse quartz syenite | >1000 | |
| | Atiyah monzogranite | Undefined | |
| | Hasha formation | | |
| | Ifal suite | | |
| | Muklar complex | | |
| | Amlas formation | 2000 | Aquitard |
| | Hinshan formation | Undefined | |
| | Zaytah formation | >1000 | Aquifuge and partially aquitard |
| | Ghawjah formation | Undefined | Aquifuge |
| | Al Marr formation | | |
| | Silasia formation | 500 | Aquitard |
| | Hegaf formation | Undefined | Aquifuge and locally Aquiclude |
| | Alkali granite | | Aquifuge |
| | Granodiorite | | |
| | Hornblende gneiss | | |
| | Gabbro and diorite | | |
| | Haql suite | | |
| | Amud formation | | |
| | Quartz monzonite | | |
| | Asmar complex | | |
| | Ghadiyah granite | | |
| | Wasit granite | | |

[a]The estimated thickness according to Clark (1987)

Sahm Sandstone rock formations occurred. These rock formations were described as members belong to Saq formation (Edgell 1997; Al-Ahmadi 2009).

In NEOM Region, the sedimentary rock sequences substantially imply sandstone which is usually described as Palaeozoic and Tertiary Sandstone (Clark 1987). This sequence exceeds 7000 m where it represent the major aquifers in NEOM Region and the surrounding. These aquifers extend from the plateau of NEOM Region to Tabuk area to the east and to Jordan in the north forming a shared aquifer, which is also named as Disi Aquifer (UN-ESCWA and BGR 2013). Hence, the aquiferous attributes and their major hydrogeological properties of the Palaeozoic Sandstone are described in the hydro-stratigraphic sequence in Fig. 6.1.

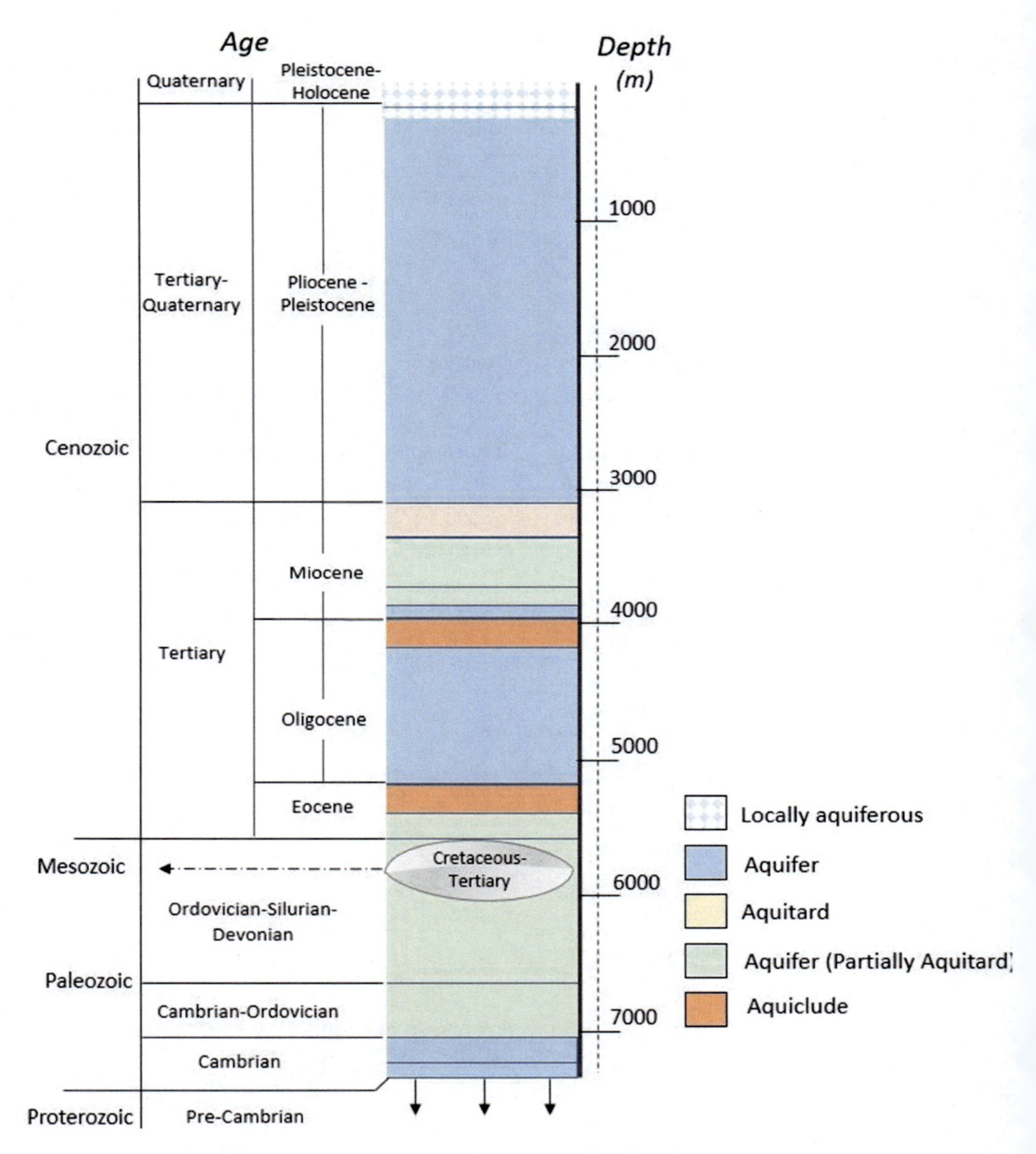


**Fig. 6.1** Hydro-stratigraphic sequence showing the aquiferous attributes and the major hydrogeological properties of the Palaeozoic Sandstone in NEOM Region

The sandstone in NEOM Region is almost porous and permeable; nevertheless, some intervenings of argillaceous and carbonate rocks occur. However, these aquifers are almost deep and, in many instances, exceed 1000 m depth.

The average depth of groundwater in the Ordovician-Silurian and Devonian Aquifer (i.e. Tabuk formation as described by Clark (1987), or Qassim formation as described by Al-Ahmadi (2009)) is 850 m. While, the average depth in the Cambrian-Ordovician Aquifer (Ram and Umm Sahm, Quweira and Siq Sandstone; which were named by Al-Ahmadi (2009) as Saq formation; and also named by UN-ESCWA and BGR in 2013 as Saq-Ram Aquifer) ranges between 650 and 1250 m. in addition, the discharge rate in the Cambrian-Ordovician Aquifer, as estimated by Al-Ahmadi (2009), is between 804 and 1728 $m^3$/day.

Whereas, another sedimentary aquifers also exist and located mainly in the pediments of the coastal zone and along the floodplains of wadis. These are mainly in alluvial deposits, sand sheets, terraces and gravel, where groundwater is found at shallow depths (50–75 m).

On the other hand, the Proterozoic rocks (i.e. basement rocks) are composed of igneous and metamorphic rocks which are of overlapping ages. These rocks are mainly of overlapping ages where rocs where they are attributed to compacted crystalline and ultramafic rocks and unassigned intrusions with abundant dikes. Therefore, they neither store groundwater nor permit groundwater to flow and thus they are mainly Aquifuge.

### *6.2.2 Groundwater Potential Map (GWPM)*

The analysis of the hydrogeology and groundwater resources in NEOM Region has been discussed in details in Sect. 6.1.1. This analysis focuses mainly on the vertical distribution of water-bearing rock formations and the other types of aquiferous rocks. Yet, the identification of potential zones for groundwater is still a challenging. In other words, it is usually tedious to identify the most suitable sites where boreholes for groundwater can be drilled. Hence, many of the drilled wells are found as wildcat wells (i.e. unsuccessful boreholes). This created debate about the creditability of the obtained studies on groundwater suitable locations.

Recently, there are new methods followed to localize the suitability for groundwater storage and even flow regime. These methods depend on the new advanced space techniques and geo-information systems. Therefore, groundwater potential maps (GWPMs) are being generated. Hence, the advantages of these methods imply: (1) identifying the potential zones for groundwater storage based on geo-spatial distribution (i.e. mappable data) (2) the produced maps are essential and supportive tools to be used when the hydrogeology of an area is not well identified, in addition (3) the applied methods are known with cheap cost, effective, less time consuming and can deal with rugged and inaccessible regions.

Similarly to the achieved study applied by the author on groundwater mapping, which was applied to Wadi Aurnah, Western Arabian Peninsula (Al Saud 2010);

therefore, the author adopted same methodology to produce the GWPM for NEOM Region where space techniques and the geo-information systems were used.

1. Concept of GWPM production

The applied approaches in groundwater exploration using remote sensing and GIS showed discrepancy and thus different results, even though, in many instances, all methods used space techniques and geo-information systems (i.e. same satellite images, software for images processing and GIS applications). The reason behind this discrepancy, as justified by the author, is attributed mainly to the differentiation of the manipulated factors acting in groundwater storage, as well as the contradicted data analysis and interpretation, notably those recognized on satellite images.

The involved factors, illustrated by many researchers, in producing GWPMs are often different and in sometimes they were selected according the physical setting of the studied areas which in turn affected the resulted maps. Therefore, it was found that some studies involved the lineaments separately in generating GWPMs (such as Teeuw 1995); others researchers added drainage systems to lineaments (such as Ahmad et al. 1984). Besides, Ganapuram et al. (2008) for example integrated the morphology, geologic structures, drainages, slopes and land cover. However, not all applied methods were successful, where as some of them showed good results. Therefore, validation of the applied methods is necessary to deduce their reliability and this can be done either on geophysical field surveys or by investigating data from drilled wells.

Other than the determination of the influencing factors on GWPMs, there is also the systematic approaches used which are significant. This represent the digital integration of these factors in the software. Therefore, the integration of the influencing factors is applied after categorizing each factor into define classes (usually five classes as adopted by the author), each class is evident to groundwater storage.

2. Materials and method

The materials and method used by the author for Wadi Aurnah Basin were carefully considered and the results were realistic enough. This has been proved by comparing the resulted map of Wadi Aurnah Basin with data from dug wells in this basin. Therefore, 2/3 of the successful water wells in the basin were located in the determined zones with groundwater potentiality.

Therefore, the adopted factors to generate GWPM for NOEM Region were slightly modified from those applied for Wadi Aurnah Basin. Thus, each factor was systematically manipulated as a geo-spatial layer. These factors are: (1) rainfall (2) lithological characteristics (3) rock fractures (4) terrain slopes (5) stream density and (6) land forms.

- Rainfall rate

Rainfall is always accounted while assessing groundwater potential zones, and it is also linked with the recharge zones on terrain surfaces (Shaban et al 2006). This is because they are (rainfall and recharge zones) considered as a source of water where groundwater fed from.

For this purpose, rainfall rate map must be prepared. In this study, it was produced depending on time series data since 1979, as previously mentioned in Chap. 3, Sect. 3.2.1 where rainfall data was adopted form several sources including: CHIRPS 2015; GAMEP 2019 and Meteoblue 2020.

Therefore, five classes for rainfall rate were adopted, and they are as follows: <15, 15–25, 25–35, 35–45 and >45 mm/year. In this respect, higher rainfall rates were always at higher altitudes and vice versa.

- Lithological characteristics

In fact, lithology is one of the most controlling factors in groundwater storage, as well as in groundwater flow. It is represented by the geo-spatial distribution of different rock formations. Thus, in generating GWPM for NEOM Region, the available geological maps of 1:250.000-scale were adopted (Rowaihy 1985; Clark 1987 and Davies and Grainger 1985).

The entire lithologies in NEOM Region can be categorized into two major pillars. They are the sedimentary sandstone and the Proterozoic basement rocks. Even though, the majority of groundwater storage is attributed to the sedimentary sandstone sequence, but the categorization of rock formations were considered all exposed rocks.

Therefore, the identified 55 rock formations and units (Table 6.1) were classified into five categories. Hence, the categorization of these 55 rock formations and units was based on the hydrogeological elements (mentioned in Sect. 6.2.1) for each formation or unit.

In addition, modifications on the existing lithologies on the available maps were performed using satellite images which were processed using ERDAD Imagine-2018, and applying the optical and digital advantages (e.g. edge detection, directional filtering) mentioned in Sect. 4.6.1. Thus, the processed images are: Landsat (30 m, 60 m thermal); Sentinel-2 (10 m, 20 SWIN); Spot-7 (1.5, 10 NIR); WV-2 (0.46 m in black and while color).

**Table 6.2** Classes of lithologies characteristics with respect to GWRP in NEOM Region

| Class | Area (km$^2$) | %[a] | Major lithology | GW potentiality |
|---|---|---|---|---|
| I | 9540 | 36 | Sandstone | Very high |
| II | 795 | 3 | Sandstone with limestone | High |
| III | 2385 | 9 | Sand and argillaceous rocks | Moderate |
| IV | 1855 | 7 | Tuffaceous rocks | Low |
| V | 11925 | 45 | Ultramafic rocks | Very low |

[a]Percentage of this class to the area of NEOM Region

Table 6.2 shows the areas of each class, where it is obvious that class-5 is the most dominant and occupies about 11925 km$^2$, 45% of NEOM Region. This evidences promising characteristics towards groundwater storage.

- Rock fractures

Normally, fractures enhance the permeability and the secondary porosity of rocks, which is in turn increases the mainly vertical water flow to recharge groundwater storage. Hence, rock fractures are often considered as one of the principal factors in the identification of groundwater potential areas as well as in groundwater flow.

The majority of fracture systems implies the presence of faults as the most acting aspects of rock structures, and usually these faults are accompanied with another aspects of rock deformations such as fissure and joints. All these structures play a role in enlarging the spacing, openings in rocks; therefore, adapt the rock to store water.

The identification of fault has been lately successfully adopted by new approaches of satellite images analysis which are supplementary to the available geologic maps. Hence, the existing fracture systems, and specifically faults, can be recognized linear features on satellite images; and these features are named as "lineaments". Thus, a lineament was defined as any observable geomorphic linear feature that attributed to geological structures (O'Leary et al. 1976).

For the purpose of faults identification was obtained from Landsat 7 ETM$^+$ and Aster images. While, ERDAS Imagine-2018 software was also used to analyse these images. Similarly to the lithological modifications, the digital applications were applied, in addition to the advantage "edge detection", thermal band application and band combination where bands 2–5–3 were ordered to show the most indicative observation of faults on Landsat images.

Therefore, all recognized geologic-related linear features from the processed satellite images were added to the faults present on the available maps, and eventually the first lineaments map for NEOM Region was produced (Fig. 6.2).

The identified lineaments on the produced map show different dimensions and orientations. However, there are several clues to deduce the relationship between faults and groundwater storage. In this view, the most applicable relationship is expressed by the author on Wadi Aurnah Basin (Al Saud 2010, 2011) where the density of lineaments. In other words, the dense lineament zones are much indicates for higher porosity and permeability and thus potential for groundwater storage.

The density of lineaments ($L_d$) was calculated by dividing the number of linear features ($L_n$), i.e. attributed to faults and fractures systems, by the area selected (A). In other words, lineament density was calculate using the formulae:

$$L_d = L_n/A$$

Therefore, NEOM Region was classified frames of define areas (5 km × 5 km; 25 km$^2$), and the number of lineaments in each frame was counted. Therefore, the number of lineaments was put in the mid-point of each frame. Consequently, a map

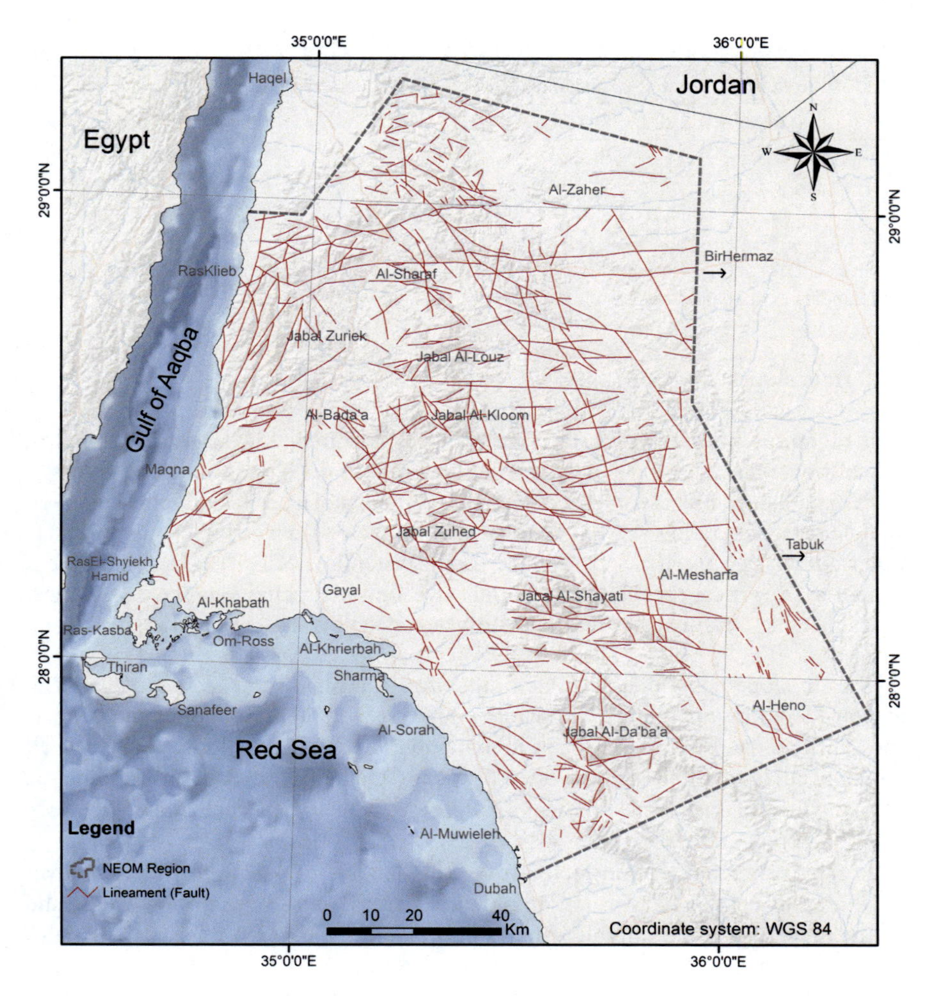


**Fig. 6.2** Lineament map of NEOM Region

with grids and nodes was resulted. From the grid map; however, contour map could be plotted representing lineaments density. The lineament density map will be used, as a geo-spatial layer for further data integration in the GIS system.

The produced map shows that the total lineaments length is about 4509 km. This map was categorized as five classes showing lineament density with respect to GWRP as shown in Table 6.3.

**Table 6.3** Lineaments density with respect to GWP in NEOM Region

| Class | Area ($km^2$) | % | Lineaments density (lineaments[a]/25 $km^2$) | GW potentiality |
|---|---|---|---|---|
| I | 992 | 22 | >40 | Very high |
| II | 641 | 14 | 40–30 | High |
| III | 1208 | 27 | 30–20 | Moderate |
| IV | 947 | 21 | 20–10 | Low |
| V | 721 | 16 | <10 | Very low |

[a]Lineaments as appear on maps with scale 1:300.000

- Terrain slopes:

While it represents a terrain surface behavior a part from the substratum characteristics; however terrain slope is involved in many studies done on groundwater storage assessment, and slope is sometime ignored in other studies, especially those applied on flat areas. According to Al Saud (2010), slope factor must be included in groundwater studies, because it has a role in increasing water flow velocity with a subsequent reduction in vertical percolation, and thus affecting the recharge processes. Hence, the higher the terrain slope the rapid surface water flow and less recharge rate and vice versa.

Terrain slope has a double-folded influence in groundwater recharge. This can be considered by applying a comparative analysis between the mountainous areas and flat lands and plains. Thus, in mountainous regions high flow energy is well pronounced and rainfall water is often flows into channels (i.e. as runoff), while the overland flow in this case is minimal. In this case, groundwater recharge will be very low (Doll et al. 2002).

In relatively gentle sloping terrains (i.e. flat lands), rainfall water is spread over large area, if compared with the mountainous areas, and overland flow exists and giving more chance for rainfall water to infiltrate into the sub-surface rock masses.

The generation of slope map has been well discussed in Chap. 4, Sect. 4.7. Hence, slope map has been done from SRTM DEM (30 m spatial resolution) by using ArcGIS software. Form the obtained map (Fig. 4.7), the similar classification used of slope, in Chap. 4, has been adopted for GWPM. This is because this classification is most reliable. Table 6.4 shows the five classes and their relationship to GWRP.

**Table 6.4** Classes of terrain slope with respect to GWP in NEOM Region

| Class | Area ($km^2$) | % | Terrain slope (degrees) | GW potentiality |
|---|---|---|---|---|
| I | 10940 | 41 | <5 | Very high |
| II | 4956 | 19 | 5–10 | High |
| III | 5274 | 20 | 10–20 | Moderate |
| IV | 3318 | 12.5 | 20–30 | Low |
| V | 2012 | 7.5 | >30 | Very low |

**Table 6.5** Stream density classes with respect to GWP in NEOM Region

| Class | Area (km$^2$) | % | Stream density (km/ km$^2$) | GW potentiality |
|---|---|---|---|---|
| I | 3180 | 12 | <5 | Very high |
| II | 4770 | 18 | 5–10 | High |
| III | 14045 | 53 | 10–20 | Moderate |
| IV | 2385 | 9 | 20–30 | Low |
| V | 2120 | 8 | >30 | Very low |

- Stream density

Streams, as the principal component in drainage systems, are significant indirect factor controlling groundwater storage. They are similar to terrain slope where they represent surficial indicators for surface water behavior and its relation to the recharge of this water to the underlying rock masses.

Drainage systems, including streams, were studied in details for NEOM Region as in Chap. 5 (Sect. 5.5) where SRTM DEM and Arc-GIS used in an integrated systematic approach. In addition, the density of these streams was calculated for each catchment separately.

However, for GWPM map the density of streams was obtained following the same method used in Chap. 5 (Sect. 5.5), but this was applied to the entire NEOM Region as a unified geographic area.

The drainage systems map, including streams, was produced from SRTM DEM and using Arc-GIS, and it was descriptively classified into four classes as coarse, medium, fine and very fine. However, for the GWPM, five streams density are required in order to have more detailed diagnose for NEOM Region. These classes and their relationship with GWP storage are shown in Table 6.5.

- Landforms

Except few studies (e.g. Scanlon et al. 2005; Al Saud 2010; Arulbalaji et al. 2019), the landform map, or sometime land cover/use map, is rarely included in the systematic generation of GWPMs, while it is very essential terrain component where the natural and man-made input on terrain surface are considered.

Groundwater is intimately connected with the landforms and land use features that it underlies, and most landforms are vulnerable to the anthropogenic activities on the land surface above. This touches the volumetric and quality of groundwater (Lerner and Harris 2009).

Therefore, landforms are intimately interrelated with the geology and geomorphology of any area, and also significantly with the land use in this area. This has a significant impact on groundwater storage where some landform components (e.g. constructions, roads, etc.) build a barrier for surface water infiltration into the substratum, while, the geology and geomorphology have also important role but with diverse impacts depending of their characteristics. Therefore, it is utmost important to involve the landform map in generating GWPMs.

**Table 6.6** Landform classes with respect to GWP in NEOM Region

| Class | Area ($km^2$) | %[a] | Landforms | GW potentiality |
|---|---|---|---|---|
| I | 5496 | 20.74 | Alluvial plain, Valley deposits, Sand sheets and hill rocks | Very high |
| II | 3692 | 13.93 | Beach sand, alluvial fans | High |
| III | 16865 | 63.64 | Coastal plain, gypseous pediplain, hills and rock outcrops, mountains | Moderate |
| IV | 344 | 1.30 | Tidal flats, Sabkha, Laval and volcanic hills | Low |
| V | 103 | 0.40 | Urban and agricultural area | Very low |

[a]The percentage of the total area of NEOM Region

As previously mentioned in Chap. 4 (Sect. 4.6.1), the production of landform map for NEOM Region was based on the available landform maps obtained by the MoA (1980). These maps were subjected to sub-setting and mosaicking. In addition, digitalized all landform components was done from the available maps using Arc-Map.

In addition to field verification and the processing of satellite images, including WV-2 and Spot-7, were undertaken to modify and update the available landform maps. Therefore, landform map for NEOM Region was produced in a unified map sheet and scale 1: 250.000 (Fig. 4.5).

For data manipulation in order to produce the GWPM and similarly to the previous obtained geo-spatial layers for the influencing factors on groundwater storage; therefore, five landform classes were adopted to represent their influence on groundwater storage. These classes are illustrated in Table 6.6.

1. Data modelling

The results maps for the five influencing factors were systematically produced in the GIS system, each factor with five-classes describing the degree of interrelationship to groundwater storage in NEOM Region. Hence, each factor encompasses five levels of effect which are from high to low effect (i.e., high GWP to low GWP).

Having all geo-spatial layers, for the influencing factors, digitally in GIS forms; therefore, then they should be systematically integrated together in order to have the final output. In other words, the geo-spatial layers would be overlapped (i.e. superimposed) in order to have one geo-spatial map sheet representing all involved factors.

However, not all the influencing factors have the same effect on groundwater storage. For example, the lineaments as a function of rock deformation has significant impact on groundwater storage which is normally large than that for stream density. Besides, rainfall rate, as the source, has the most significance role. Moreover, some factors negatively act in the enhancement of potentiality of groundwater storage, such as terrain slope factor, but this is the opposite for the dense lineaments and so on. For this reason, each factor was given a specific weight of effect on ground

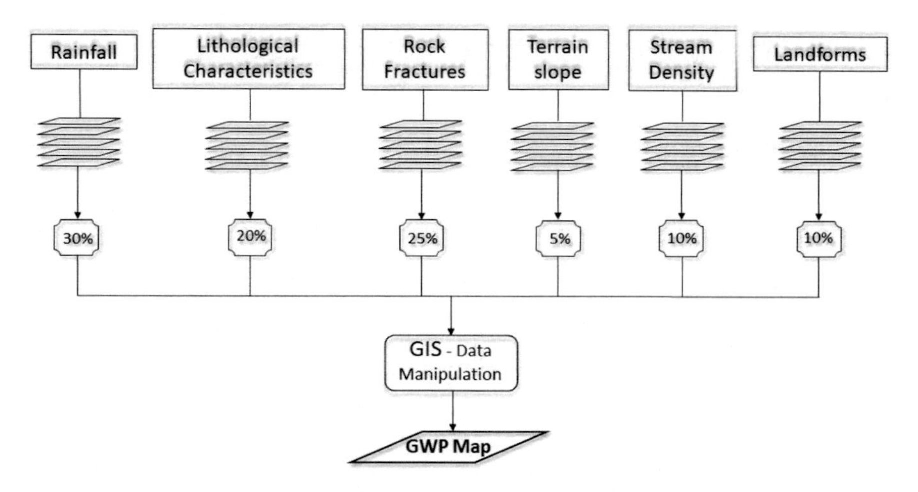


**Fig. 6.3** Model chart showing the factors used to generate the GWPM for NEOM Region

water storage. This has been lately adopted in several studies (e.g. Al Saud 2010; Chaudhary and Kumar 2018; Andualem and Demeke 2019).

The geo-spatial layers for each influencing factor on groundwater storage was given the following weights (Fig. 6.3):

- Rainfall = 30%
- Lithological characteristics = 20%
- Rock fractures = 25%
- Terrain slope = 5%
- Stream density = 10%
- Landforms = 10%

The adopted weights for these factors in were determined according to the expertise from field observations as well as they were also adapted from several obtained studies on the subject matter (e.g. Edet et al. 1998; Robinson et al. 1999; Shaban et al. 2006; Chaudhary and Kumar 2018).

For detailed assessment of the influencing factors, rates are also given as sub-scores for the adopted weights. Hence, the rates were classified between (I) for very high potential rate and (V) for very low potential rate (Table 6.6).

If the maximum value for each rate is 100%, therefore, the rates will be classified according to the following ranges: 100–80%, 80–60%, 60–40%, 40–20% and 20–0%. This means that the average of each rare range will be: 90, 70, 50, 30 and 10% for ranges from I to V; respectively (Table 6.6).

For calculating the effectiveness for each of the influencing factors on groundwater storage (FE), the weight ($F_W$) and rate ($F_R$) variables must be accounted. Consequently, weights are multiplied by rates as follow:

$$FE = F_w \times F_r$$

For example, the weight of class (III) for the lithology factor equals 50%, if multiplied by the factor weight, which is 20; therefore, the factor effectiveness will be:

$$\text{EF} = 50/100 \times 20 = 10$$

While for the landforms factor, for example, the factor effectiveness for class (II) will be:

$$\text{EF} = 70/100 \times 10 = 7$$

Therefore, class (I) in the rainfall factor is the most effect on groundwater storage where it was calculated at 27, whereas the least one is class V in the slope factor which is 0.5.

ESRI's Arc View software was used for factors manipulation including the systematic calculations for their weights and rates, and thus in the overlapping of the geo-spatial layers. In this respect, the 5-classes factors were manipulated systematically in a map with 4 GWP zones in order to be more realistic in the assessment of the optimal GWPM.

The produced map for GWP for NEOM Region were represented by a number of polygons each evidences potentiality to groundwater storage (Fig. 6.4). Hence, the obtained for GWP zones on the produced map was described as: high potentiality, moderate, low and uncertain potentiality for groundwater storage. Thus, the zone attributed to "uncertain potentiality" was considered in order to assure that these mountainous regions in the "uncertain zone" need in-depth investigation notably that the largest part of exposed rocks are of lithological complexes (mainly basement rocks).

Based on field validation and the collected data and info from the accessible boreholes in NEOM Region, it was obvious that there is a clear consistent between the identified GWP zones and groundwater depths as follows:

- High GWP: average depth >800 m, area 1625 $km^2$
- Moderate GWP: average depth >75 m, area 902 $km^2$
- Low GWP: average depth >350 m, area 382 $km^2$
- Uncertain GWP: un defined depth, area 23591 $km^2$.

## 6.3 Ore Deposits

Rock contain considerable amount of minerals are called an ores. Thus if these ores contain profitable mineral deposit, then they are called ore deposits, and these ore deposits can be either metallic or non-metallic minerals.

Therefore, metallic ores, as the opposite of non-metallic ores, are composed mainly of metals in their original form, and they characterize by the following:

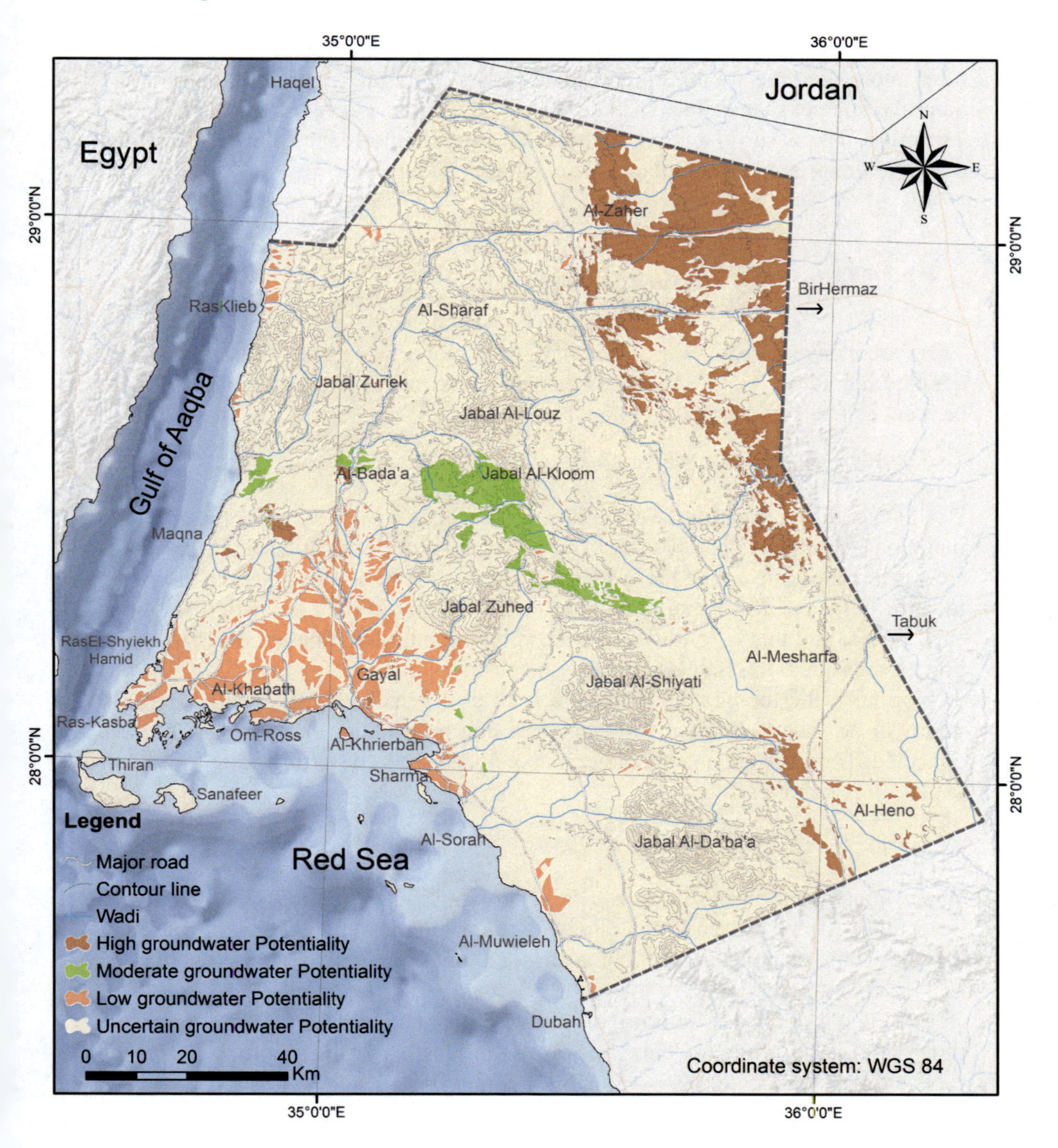


**Fig. 6.4** Groundwater potential map of NEOM Region

- They are mainly found in igneous and metamorphic rocks.
- They can be make as new product by on melting.
- Good conductors of heat and electricity
- They are malleable and ductile.
- They have luster.

Natural resources makes up a significant proportion of the wealth of many nations, often more than the wealth embodied in produced capital; therefore, considering natural resources management a major factor of economic development (WB 2006). Therefore, it is not surprisingly, countries richly with natural capital have the potential to derive important current income from natural resources.

### 6.3.1 *Methods of Identification*

There are many approaches used identify and explore ore deposits stating from very detailed laboratory testing to comprehensive observation from space tools. These element can be summarized as follows:

1. Geological methods:

Usually field investigation is primarily performed, and this is accompanied with mapping techniques. Therefore, field mapping and rock verification and sampling as well as trenching are applied in order to deduce key elements of a potential discovery at surface. Hence, rocks and mineral at tested directly in the filed using simplified approaches, and this is followed by recording the interpretations in order to create precise measurements and high-quality maps on a small scale.

Normally, the topographic maps are used at first as base maps, and a geology map will be produced when sufficient data and information on the area has been collected. Thus, the obtained geology map will be used as a starting point for further exploration activities.

2. Stripping and trenching:

When mineralization is recognized in rock exposures, sampling will be done to determine its physical and mineralogical characteristics and accurately map it is locality. Thus, stripping and trenching are sometimes required to remove overburden rocks and materials.

Stripping is applied by removal of shallow amount of surficial materials (e.g. soil, dirt, gravel, etc.), while trenching makes use of blasting or digging to each into the rock and uncover a columnar section of mineralization.

3. Geochemical testing:

The collected samples from an exploration locality will be testing in the laboratory where these sample are prepared according to the requirement of each test, such as drying, crushing and/or milling; to determine the chemical elements that make up the sample

There is a wide variety of testing methods. For example, prepared sample is to dissolve in an acid mixture and analyse it using an Inductively Coupled Plasma (ICP) device that can measure the relative mass of each element using a technique called

Mass Spectrometry (MS); as well as addition X-Ray diffraction is applied to analyse the mineralogical contents of the rock sample. In addition, mineralogical analysis is also applied using microscopes, for example, in optical mineralogy analysis to identify the mineralogical characteristics including texture and crystallography.

4. Geophysical techniques:

Geophysical techniques are a set of analytical approaches that identify the physical properties of minerals and rocks, including:

- Magnetism—Measuring in the amount of magnetite/pyrrhotite in rock samples; typically used to find volcanogenic massive sulphide (VMS) deposits and kimberlites.
- Electromagnetic (EM)—It is usually used to detect metals and deposits hosting sulphide minerals of interest.
- Electrical sounding—It is applied mainly for conducting minerals in rocks and sediments.
- Induced polarization (IP)—A method which measures the response of the substratum after propagating an electrical current. It is useful in exploring, for example, certain porphyry copper, VMS and precious metal vein deposits.
- Radioactivity detection—Special devices (e.g. Geiger counter and scintillometers) are used to detect radioactive elements, such as uranium, thorium, etc.

### *6.3.2 Ores Exploration in Saudi Arabia*

Saudi Arabia encompasses two geological provinces with marked difference in age and lithological characteristics. First, is the Arabian Shield, a major source for mineral occurrences, located in western part of Arabian Peninsula and it is composed of the Precambrian metamorphosed rocks. While, the second province comprises younger rock sequences of the Paleozoic, Mesozoic and Tertiary rocks.

The Arabian Shield, where NEOM Region is an extension of it, occupies tremendous metallic mineral finds, and encompasses magmatic deposits within the igneous and metamorphic rocks. In addition, non-metallic resources are widespread all over the Kingdom including the Arabian Shield. Therefore, Kingdom of Saudi Arabia promises a suitable environment which is potentially rich in variety of metallic and non-metallic mineralization (Darwish and Butt 1996).

There are more than 7100 mineral occurrences have been discovered in the Kingdom of Saudi Arabia Nehlig et al. (1999). In 1970, the database of all exploration drill-holes for these occurrences were set up as computerized data, by BRGM (Bureau de Recherches Géologiques et Minières) for the interest of Deputy Ministry for Mineral Resources (DMMR), and then Drill Hole Information System (DHIS) was created. This has been developed and updated to Mineral Occurrence Documentation System (MODS) which recorded all mineral occurrences discovered in the Kingdom.

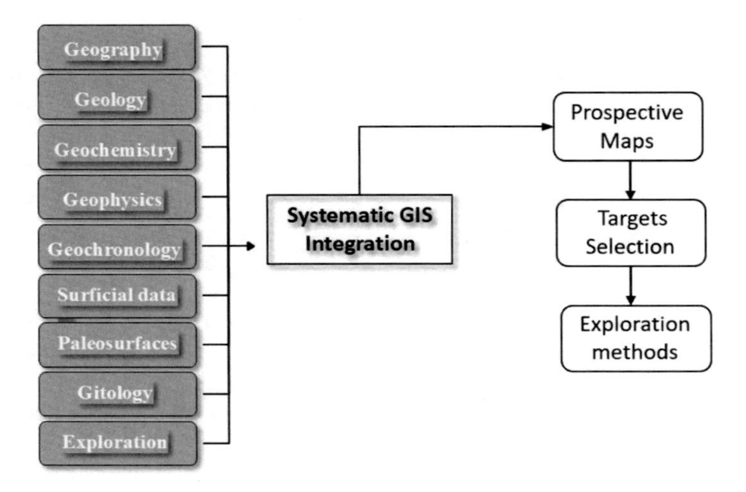


**Fig. 6.5** Major integrated elements of Metallogenic data reassessment in the GIS system as performed by DMMR, SGS, BRGM and ESRI

Therefore, all recognized minerals in Saudi Arabia have recorded in the MODS database by DMMR and the Saudi Geological Survey (SGS). Each site was assigned for a serial number, with a detailed description (e.g. location, type of mineralized body, condition of discovery, chemical analyses, etc.).

Lately, geological and mineral exploration in the Arabian Shield have been well pronounced. Thus, it was reported that there two mines in exploitation, many mining projects at the feasibility or investment stage, plus several promising occurrences at the evaluation stage (Nehlig et al. 1999).

Due to the reason that most results and data were heterogeneous and has rarely been synthesized, as well as maps were available only as hard copies. Therefore, DMMR as well as the SGS, BRGM and ESRI performed a program for data and information re-evaluation and compilation under using GIS systematic and modeling approaches. These include geography, geology, structures, geochemistry, aeromagnetic, geochronology, paleo-surfaces, and surficial formations which are linked to a mineralized-occurrence database. This made reassessment and organization of the metallogenic data and quantitative measurement of the regional- and local-scale geological, geophysical and geochemical exploration (Fig. 6.5).

### *6.3.3 Ores Exploration in NEOM Region*

There is no specific study on the economic ore deposits dedicated for NEOM Region, but as previously mentioned, there many surveys obtained for the entire Saudi Arabia including the Arabian Shield and Tabuk Province where NEOM Region is located. Therefore, the author aimed at making an assessment to organize and highlight on the economic ores deposits in NEOM Region whether these deposits are metallic or non-metallic ores.

Therefore, this assessment was based mainly on the obtained studies and documents (e.g. USGS-ARAMCO 1963; Béziat et al. 1995; Nehlig et al. 1999) as well as the geologic maps done by Clark (1987) and Davies and Grainger (1985) for Al-Bada'a map (sheet: 28 A) and Al-Muieleh map (sheet: 27 A); respectively. Thus, the author studied all available documents and investigated the lithological characteristics of the existing rock formation where field surveys and geological measurements have been carried out.

As previously mentioned that Neom Region is one of the 18 mineralized districts in the Saudi Arabia Nehlig et al. (1999). Hence, many studies and survey campaigns have been done to this region within the context of the exploration ore deposits in the entire Saudi Arabia with a focus on the region of Arabian Shield (Béziat et al. 1995). Therefore, NEOM Region occupies at least 25 major economic ore deposits, while tremendous associated minerals are also exist.

The most potential economic metallic ores in NEOM Region are those located in the mineralized Jabal At-Tawileh microgranite and the jaspilitic iron ore related to the Wadi Sawawin deposits. While, base metal occurrences are widespread; specifically in the Tertiary rock formations (Clark 1987). Therefore, the explored metallic ores in NEOM Region are:

Silver (Ag), gold (Au), beryllium (Be), chromium (Cr), copper (Cu), iron ores (hematite, magnetite, jaspilitic), lanthanum (La), manganese (Mn), molybdenum (Mo), niobium (Nb), nickel (Ni), lead (Pb), rare earth elements (REE), sulfur (S), tin (Sn), strontium (Sr), Zinc (Zn), and titanium (Ti).

For the non-metallic ores, the large amount of gypsum-anhydrite of the Maqna massif is the most economic ores in NEOM Region. While, industrial minerals, such a silica sand, grave, building stone and marble are available locally (Clark 1987). Therefore, the explored non-metallic ore in NEOM Region are:

Barite ($BaSO_4$), fluoride (F), marble, nepheline (Ne), phosphorite ($P_2O_5$) silica sand ($SiO_2$), gypsum-anhydrite ($CaSO_4$).

The majority of geographic distribution of metallic ores is mainly in the mountainous ridges, while the non-metallic ore are distributed mainly in the plateau and the coastal zone. Besides, the geologic distribution of these minerals was summarized in Table 6.7.

The geographic distribution of these ores are lithological and structurally controlled. Examples of these distributions for metallic and no-metallic ores are shown in Figs. 6.6 and 6.7.

According to DMMR, as mentioned in Clark (1987), the geographic distribution of these ores can be summarized as follows:

1. Maqna area: ($CaSO_4$, S, $P_2O_5$, Sr, $BaSO_4$, Cu, Zn, Pb, Ag, F).
2. Jabal Zuhed: (iron ores, Cu, Zn, Pb, Mo, Ag, Sr, Ti, Ni, Nb, REE, Sn, Be, La, F).
3. Jabal Al-Shayati: (Cu, Zn, Mo, iron ores, $SiO_2$).
4. Tayyeb Al Ism: (Mn, iron ores, Ti, Cu, Ne, Mo).

**Table 6.7** Existed economic ore deposits (metallic and non-metallic) in NEO Region

| Period | Rock formation | Existed ores[a] |
|---|---|---|
| Q | Undifferentiated deposits of sand, gravel, Aeolian, alluviums and Sabkha | Silica sand |
| T-Q | Conglomerates | – |
| | Lisan formation | – |
| T | Bad formation | Barite, Sulfer |
| | Nutaysh formation | Copper, Gypsum-Anhydrite |
| | Musayr formation | Copper, Lead, Barite, Phosphorite |
| | Usayliyah formation | Marble |
| | Khuraybah formation | Copper, Lead, Gypsum-Anhydrite |
| | Sharik formation | – |
| | Azlam formation | Copper, Zinc, Lead |
| | Adaffa formation | – |
| C-T | Gabbro and diorite dikes | – |
| O-S-D | Tabuk formation | – |
| Cmb-O | Ram and Umm Sahm sandstones | – |
| Cmb | Quweira sandstone | – |
| | Siq sandstone | Silica sand |
| | Sawda complex | Nepheline |
| Pre-Cmb | Undifferentiated volcanics | Molybdenum, Gold, Copper, Zinc, Silver |
| | Lawz complex | – |
| | Minaweh formation | – |
| | Sulaysiyah formation | – |
| | Saluwah formation | – |
| | Shar complex | – |
| | Dabbagh complex | – |
| | Massah complex | Marble |
| | Mabrak granite | Marble |
| | Maharish complex | – |
| | Sadr complex | – |
| | Sawawin complex | Iron ore, Manganese, Marble, Beryllium, Strontium, Chromium, Mable |
| | Duba complex | – |
| | Midyan suite | Nepheline, Lanthanum, Zirconium, Copper, Lead, Marble |
| | Mowasse quartz syenite | REE |
| | Atiyah monzogranite | Fluoride, Molybdenum, Phosphorite, Copper, Barite |

(continued)

**Table 6.7** (continued)

| Period | Rock formation | Existed ores[a] |
|---|---|---|
| | Hasha formation | Silica sand |
| | Ifal suite | Nickel, Silver, Molybdenum, Zinc, Fluoride, Copper, Iron ore |
| | Muklar complex | – |
| | Amlas formation | Molybdenum, Copper, Iron ore |
| | Hinshan formation | – |
| | Zaytah formation | Marble |
| | Ghawjah formation | Gold, Copper |
| | Al Marr formation | Gold, Copper |
| | Silasia formation | Iron ore |
| | Hegaf formation | Iron ore, Niobium, Titanium, Copper, Marble |
| | Alkali granite | – |
| | Granodiorite | – |
| | Hornblende gneiss | – |
| | Gabbro and diorite | – |
| | Haql suite | REE, Beryllium, Tin, Marble |
| | Amud formation | – |
| | Quartz monzonite | – |
| | Asmar complex | – |
| | Ghadiyah granite | – |
| | Wasit granite | Marble |

*Q* quaternary, *T* tertiary, *C* cretaceous, *D* Devonian, *S* Silurian, *O* Ordovician, Cmb Cambrian, *pre-Cmb* pre-Cambrian

5. Jabal Al-Louz: (iron ores, Cu, REE, Mo, Be, Ni, F, Cr, La, Zn, Ti)
6. Wadi As-Sahab: (Marble, $SiO_2$).
7. Wadi Rawa'a: (Ni) in fold core.
8. Wadi Hawja: (Cu) in conglomerate.
9. Wadi Habib: (Mo, F) associated with dikes.
10. Wadi Amlas (iron ores).

No qualitative estimates are available except for the iron ores which were classified, according to British Steel Cooperation (1981), as follows:

1. Wadi Rawa'a: 32.4 million tones (36–47%).
2. Sharma: 9.8–15.8 million tones (38%).
3. Aynouna: 4.5 million tones (30%).
4. Jabal Abou Judaydat: 2.3 million tones (34–36%).
5. As-Shrigri: 92 million tones (38–40%).

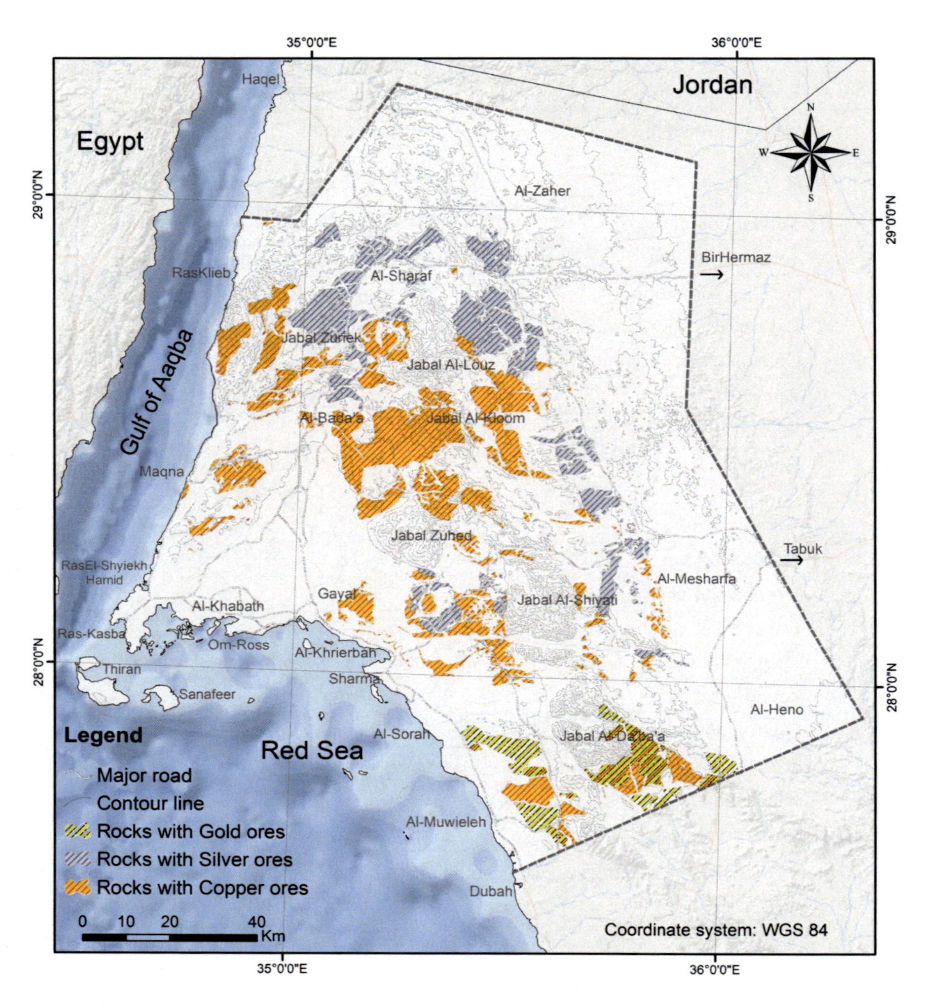


**Fig. 6.6** Examples for the distribution of rock bodies with metallic ore in NEOM Region

Minor mineral occurrences were found in basement rocks associated with either intrusive or contact zones or quartz veins; in places mineralization is present along faults and fractures. Thus, BRGM investigated Zn–Cu prospect in Wadi Sharma in 1979. While, Mn, iron, Ti, Cu, Zn, and Mo were reported in Jabal Sabil, at Siq and near As–Sadd, where 10.000 ppm Cu, 500 ppm Mo and 200 ppm Zn (Trent and Johnson 1966).

Mineralization related mainly to quartz veins and dikes is present near Jabl Al Abyad, NE of Sharma where a load of magnetite and traces of Ag, estimated at 75 to 100 tones, were explored (Burton 1878).

The Tertiary mineralization showed base metals that are commonly associated with barite and occur SW to the area of study. While, much Copper-related Zinc

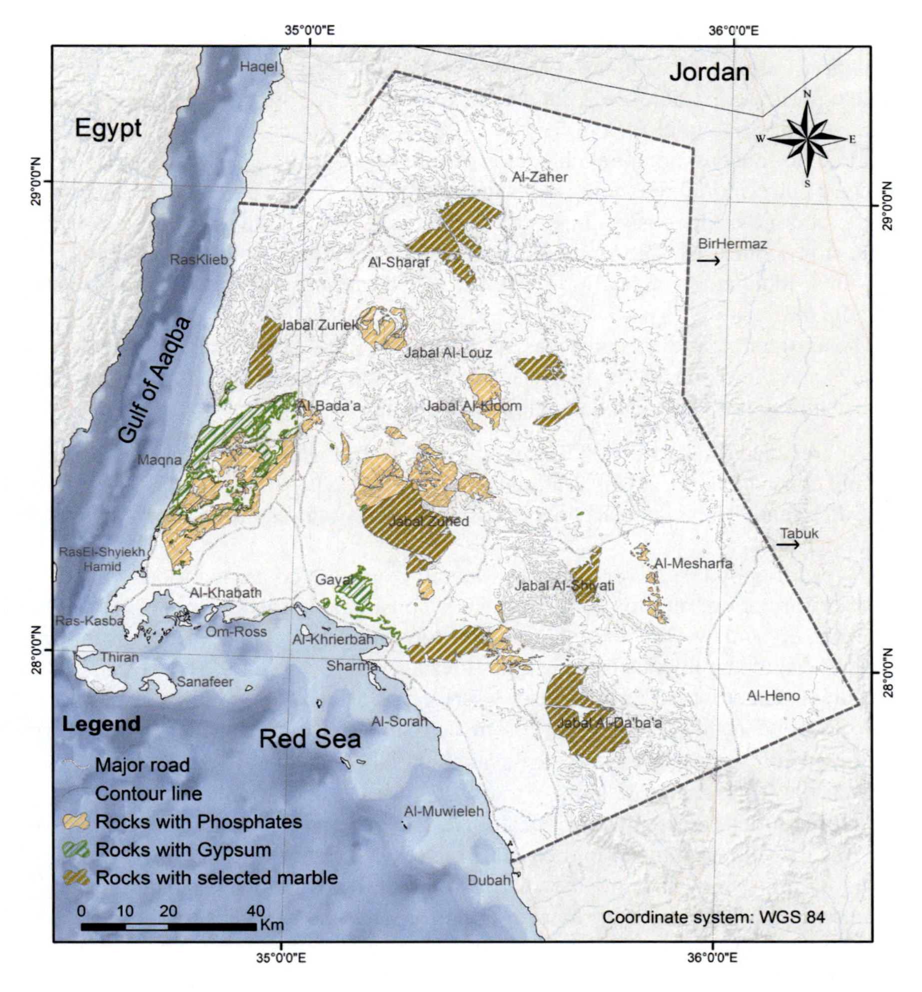


**Fig. 6.7** Examples for the distribution of rock bodies with non-metallic ore in NEOM Region

mineralization probably originated by syn-deposition associated with faulting (Motti et al. 1982).

Also, Tertiary mineralization showed large amounts of pure gypsum and anhydrite that occur in Maqna massif where accessory of halite (NaCl) is present at several horizons, but it is impure. In addition, sulfur is also found associated with gypsum and anhydrite in some localities where Jabal Kibrit was estimated to contain an amount of sulfur ranges between 3000 and 4000 t (Bouge 1953).

Low-grade phosphates is another significant Tertiary mineralization ores found in Maqna massif. These are associated with glauconitic sandstone, such as in Jabal Hamada where three layers of phosphate (about 0.5 m thick) occur. The grade of $P_2O_5$ ranges between 7.75 and 22.75% (Vial et al. 1983).

Out of the 53 recognized rock formations in NEOM Region, there are 27 rock formation unexplored or unassigned for reconnaissance and testing as shown in Table 6.8. However, there are 21 of the unexplored rock formations with lithological characteristics potential for ore bearing. While, only 5 rock formations are neither explored nor contain potential characteristics to occupy economic ore deposits.

The 21 unexplored rock formations, in NEOM Region, are potential to contain 13 type economic ore deposits, these are: aggregates, Silica sand, gypsum-anhydrite, dimension stones, Marble, Uranium, radioactive elements (other than uranium), iron ores, Manganese, Beryllium, Strontium, Chromium and Nepheline as shown in Table 6.8.

The assignment of these ores was based on their lithological characteristics, field surveys and the comparative analysis with the existing ore in rock formations with similar lithologies. This can be summarized as follows:

1. Aggregates: These are well observed in wadis and alluvial fans where alluvium and colluvial deposits are common.
2. Silica sand: These deposits are common in the existed sand formations with least impurities.
3. Gypsum-anhydrite: Even these non-metallic ores are contained locally in Lisan rock formation, but this formation has a considerable thickness (i.e. exceeding 3000 m) and wide geographic extent which gives a chance for gypsum and anhydrite to be in considerable quantities as well.
4. Dimension stones: They are also described as decoration stones, thus all rocks with good rigidity, color and low resistivity for weathering can be used are dimension stones, and these area available in the area of study.
5. Marble: This is almost similar with dimension stones, but rocks selected for marble industry are often with minimal fissures and can produce rock panels. This is well pronounced in rock mass of NEOM Region.
6. Uranium: It is well know the shallow marine deposits of phosphate is usually contain uranium and thorium ores (Ghadeer et al. 2019). This characterizes the two mentioned rock formations in Table 6.8.
7. Radioactive elements: This may include many variety of radioactive elements (e.g. U, V, Th, Ra, etc.). Hence, the presence of sandstone, which is an environment for uranium ores, with interbedded reducing horizons such as marl and evaportie rocks is a catalyst to deposit radioactive elements.
8. Metallic minerals, such as those mentioned in Table 6.8 (Iron ores, Mn, Be, Sr and Cr), were assigned as potential for Sadr Complex. This was based on the lithological similarity between Sadr Complex and other rock formations containing these minerals.

**Table 6.8** Potential rock formations for economic ore deposits in NEOM Region

| Period | Rock formation | Potential ores |
|---|---|---|
| Q | Undifferentiated deposits of sand, gravel, Aeolian, alluviums and Sabkha | Aggregates |
| T-Q | Conglomerates[a] | – |
| | Lisan formation | Silica sand, Gypsum-anhydrite |
| T | Al-Bada'a formation | Uranium, Thorium |
| | Nutaysh formation | – |
| | Musayr formation | – |
| | Usayliyah formation | – |
| | Khuraybah formation | – |
| | Sharik formation | – |
| | Azlam formation[a] | – |
| | Adaffa formation | Silica sand, Uranium, Thorium |
| C-T | Gabbro and diorite dikes | – |
| O-S-D | Tabuk formation | Silica sand |
| Cmb-O | Ram and Umm Sahm Sandstones | Silica sand |
| Cmb | Quweira sandstone | Radioactive elements |
| | Siq Sandstone | Radioactive elements |
| | Sawda complex | – |
| Pre-Cmb | Undifferentiated volcanics | – |
| | Lawz complex | Marble |
| | Minaweh formation[a] | |
| | Sulaysiyah formation | Radioactive elements |
| | Saluwah formation | Radioactive elements |
| | Shar complex | Marble |
| | Dabbagh complex | Marble |
| | Massah complex | – |
| | Mabrak granite | – |
| | Maharish complex | Marble |
| | Sadr complex | Iron ores, Manganese, Beryllium, Strontium, Chromium, Marble |
| | Sawawin complex | – |
| | Duba complex | Marble |
| | Midyan suite | |
| | Mowasse quartz syenite | Nepheline |
| | Atiyah monzogranite | – |

(continued)

**Table 6.8** (continued)

| Period | Rock formation | Potential ores |
|---|---|---|
| | Hasha formation | – |
| | Ifal suite | – |
| | Muklar complex[a] | – |
| | Amlas formation | – |
| | Hinshan formation | Dimension stones |
| | Zaytah formation | – |
| | Ghawjah formation | – |
| | Al Marr formation | – |
| | Silasia formation | Dimension stones |
| | Hegaf formation | – |
| | Alkali granite | Marble |
| | Granodiorite | Marble |
| | Hornblende gneiss | Marble |
| | Gabbro and diorite | Aggregate, dimension stones |
| | Haql suite | – |
| | Amud formation[a] | – |
| | Quartz monzonite | Marble |
| | Asmar complex | Marble |
| | Ghadiyah granite | Marble |
| | Wasit granite | – |

[a]Rock formation with no ore reported or expected
*Q* quaternary, *T* tertiary, *C* cretaceous, *D* Devonian, *S* Silurian, *O* Ordovician, *Cmb* Cambrian, *Pre-Cmb* pre-Cambrain

## References

AGI (American Geological Institute). (1980). *Glossary of geology: Falls Church.* Virginia: American Geological Institute.

Agocs, W., & Keller, F. (1962). *Airborne magnetometer-scintillation-counter: Preliminary Report.* Saudi Arabian Directorate General of Mineral Resources. Open-file Report DGMR-141, 13 pp.

Ahmed, S. (1972). Geology and petroleum prospects in eastern Red Sea. *American Association of Petroleum Geologists Bulletin, 56*(4), 707–719.

Ahmed, F., Andrawis, A., & Hagaz, Y. (1984). Landsat model for groundwater exploration in the Nuba Mountains, Sudan. *Advances in Space Research, 4*(11), 123–131.

Al Saud, M. (2010). Mapping potential areas for groundwater storage in Wadi Aurnah Basin, western Arabian Peninsula, using remote sensing and geographic information system techniques. *Hydrogeology Journal, 18,* 1481–1495.

Al Saud, M. (2011). Use of satellite images for mapping and analyze the fracture systems in western regions of Saudi Arabia. In *9th Geological Symposium for the Saudi organization for Earth Sciences* (pp. 26–28). King Saud University.

Al-Ahmadi, M. (2009). Hydrogeology of the Saq AquiIron orer Northwest of Tabuk, Northern Saudi Arabia. *JKAU: Earth Sciences, 20*(1), 51–66.

Almutairi, K., Al-Shami, S., Khorshid, Z., & Moawe, M. (2016). Floristic diversity of Tabuk province, north Saudi Arabia. *The Journal of Animal and Plant Sciences, 26*(4), 1019–1025.

Aloufi, A. A., & Amr, Z. S. (2018). Carnivores of the Tabuk Province, Saudi Arabia (Carnivora: Canidae, Felidae, Hyaenidae, Mustelidae). *Lynx, new series, 49*(1), 77–90. https://doi.org/10.2478/lynx-2018-0010.

Andualem, T., & Demeke, G. (2019). Groundwater potential assessment using GIS and remote sensing: A case study of Guna tana landscape, upper blue Nile Basin, Ethiopia. *Journal of Hydrology: Regional Studies, 24*, 100610. https://doi.org/10.1016/j.ejrh.2019.100610.

Arulbalaji, P., Padmalal, D., & Sreelash, K. (2019). GIS and AHP techniques based delineation of groundwater potential zones: A case study from Southern Western Ghats, India. *Scientific Reports, 9,* 2082. https://doi.org/10.1038/s41598-019-38567-x.

ASCE (American Society of Civil Engineers). (1985). Manual 40—Ground water management.

Béziat, P., Bache, J., Tawfiq, M., Cottard, F., Abdulhay, G., Iron orelenc, J., ... Caïa, J. (1995). *Metallic mineral deposits map of the Arabian Shield, Kingdom of Saudi Arabia, Scale: 1:1,000,000.* Jiddah-Orléans: DMMR-BRGM joint publication, two sheets.

Bouge, R. (1953). *Geological Reconnaissance of northwest Saudi Arabia: Saudi Arabia Directorate General of Mineral Resources*, Open-file Report, BRGM-OF-27, 31 pp.

British Steel Cooperation (Overseas Services). (1981). *Exploitation of Sawawin ore deposits* (Vol. 2. Geology: 289 pp).

Burton, R. (1878). *The gold mines of Midian and the ruined Midianite cities* (1979 Ed.): London, C, Keegan Paul, 392p.

Chaudhary, B., & Kumar, S. (2018). Identification of Groundwater Potential Zones using Remote Sensing and GIS of K-J Watershed, India. *Journal of the Geological Society of India, 91*(6), 717–721. https://doi.org/10.1007/s12594-018-0929-3.

CHIRPS. (2015). Climate Hazards Group InfraRed Precipitation with Station data. Available at: http://chg.geog.ucsb.edu/data/chirps/.

Clark, M. (1987). Geologic map of Al-Bada'a quadrangle, A-28; (1:250.000). Ministry of Petroleum and Mineral Resources.

Colleman, R. (1977). Geologic background of the Red Sea, in Red Sea Research 1970–1975. *Saudi Arabian Directorate General of Mineral Resources. Bulletin,22,* C1–C9.

Darwish, M., & Butt, N. (1996). Mineral resources potential and its development in Saudi Arabia. *Journal of King Abdulaziz University: Engineering Science, 8,* 107–120.

Davies, F., & Grainger, D. (1985). Geologic map of Al-Muieleh quadrangle, A-27; (1:250.000). *Ministry of Petroleum and Mineral Resources.*

Doll, P., Lehner, B., & Kaspar, F. (2002) Global modeling of groundwater recharge. In *Proceedings of 3rd International ConIron orerence on Water Resources and the Enironment Research* (Vol. 1, pp. 27–33). Germany: Technical University of Dreseden.

Edet, A., Okereke, S., Teme, C., & Esu, O. (1998). Application of remote sensing data to groundwater exploration: a case study of the Cross River State, southeastern Nigeria. *Hydrogeology Journal, 6*(3), 394–404.

Edgell, H. (1997). AquiIron orers of Saudi Arabia and their geological framework. *The Arabian Journal of Science and Engineering, 22*(1c), 5–31.

GAMEP (The General Authority of Meteorology and Environmental Protection). (2019). *Tabuk climatic data.* Available at: https://www.pme.gov.sa/Ar/Meteorology/Pages/ClimateReport.aspx.

Ganapuram, S., Kumar, G., Krishna, I., Kahya, E., & Demirel, M. (2008). Mapping of groundwater potential zones in the Musi basin using remote sensing and GIS. *Advances in Engineering Software, 40*(7), 506–518.

Ghadeer, A., Ibrahim, A., & Al-Masri, M. (2019). Geochemistry of uranium and thorium in phosphate deposits at the Syrian coastal area (Al-Haffah and Al-Qaradaha) and their environmental impacts. *Environmental Geochemistry and Health, 41*(5), 1861–1873. https://doi.org/10.1007/s10653-018-0221-x.

Ivestopedia. (2020). Ten Countries With The Most Natural Resources. Available at: https://www.investopedia.com/articles/markets-economy/090516/10-countries-most-natural-resources.asp.
Lerner, D., & Harris, R. (2009). The relationship between land use and groundwater resources and quality. *Land Use Policy, 26,* S265–S273. https://doi.org/10.1016/j.landusepol.2009.09.005.
Lujala, P. (2003). Classification of natural resources. In *The 2003 ECPR Joint Session of Workshops.* 31 March, 2003. Edinburgh, UK. 21 pp.
Meteoblue. (2020). Climate Tabuk. Available at: https://www.meteoblue.com/en/weather/historyclimate/climatemodelled/tabuk_saudi-arabia_101628.
Motti, E., Teixido, L., Vazquez-Lopaz, R., & Bigot, M. (1982). *Maqna massive area: geology and mineralization: Saudi Arabia Deputy Ministry of Mineral Resources*, Open-file Report, BRGM-OF-02-16, 44 pp.
Nehlig, P., Salpeteur, I., Asfirane, F., Bouchot V., Eberlé J., Genna A., … Tourlière, B. (1999). *The Mineral Potential of the Arabian Shield: A reassessment. IUGS/UNESCO Deposits Modeling Workshop.* BRGM. USGS—Special Publication. November 1999. https://doi.org/10.13140/rg.2.1.1889.9282.
O'Leary, D., Friedman, J., & Poh, H. (1976). Lineaments, linear, lineations: some standards for old terms. *Geological Society of America Bulletin, 87,* 1463–1469.
Robinson, C., El-Baz, F., & Singhory, V. (1999). Subsurface imaging by RADARSAT: Comparison with Landsat TM data and implications for groundwater in the Selima area, northwestern Sudan. *Remote Sensing Abstracts, 25*(3), 45–76.
Rowailhy, M. (1985). Geologic map of Haqel quadrangle, A-29; (1:250.000). *Ministry of Petroleum and Mineral Resources.*
Scanlon, B., Reedy, R., Stonestrom, D., Prudiczand, K., & Dennehy, (2005). Impact of land use and land cover change on groundwater recharge and quality in the southwestern US. *Global Change Biology, 2005*(11), 1577–1593. https://doi.org/10.1111/j.1365-2486.2005.01026.x.
SERI & GLOBAL. (2009). Overconsumption? Our use of the world´s natural resources. Available at: http://www.foe.co.uk/sites/default/files/downloads/overconsumption.
Shaban, A., Khawlie, M., & Abdallah, C. (2006). Use of remote sensing and GIS to determine recharge potential zones: the case of occidental Lebanon. *Hydrogeology Journal, 14*(4), 433–443.
Teeuw, R. (1995). Groundwater exploration using remote sensing and a low-cost geographic information system. *Hydrogeology Journal, 3*(3), 21–30.
Trent, V., & Johnson, R. (1966). Reconnaissance mineral and geological investigation in the Al Al-Bada'a quadrangle, Aqaba area. Saudi Arabia: US Geological Survey Technical letter TL-50, 18P.
UN-ESCWA & BGR (United Nations Economic and Social Commission for Western Asia; Bundesanstalt für Geowissenschaften und Rohstoflron ore). 2013. Inventory of Shared Water Resources in Western Asia. Beirut.
USGS-ARAMCO. U. S. Geological Survey-Arabian American Oil Company. (1963). Geologic map of the Arabian Peninsula: U.S. Geological Survey Misc. Inv. Map I-270 A, scale 1:2,000,000.
Venables, A. (2016). Using natural resources for development: Why has it proven so difficult? *Journal of Economic Perspectives, 30*(1), 161–184. https://doi.org/10.1257/jep.30.1.161.
Vial, A., Clintzboekel, C., Al Sati, R. (1983). *Phosphate prospecting in the Upper Cretaceous, Eocene and Miocene of the Red Sea coastal zone.* Saudi Arabia Deputy Ministry of Mineral Resources, Open-file Report, BRGM-OF-03-326, 41 pp.
WB (World Bank). (2006). *Where is the wealth of nations? Measuring capital for the 21st century.* Washington DC: World Bank.
WMO (World Meteorological Organization). (1974). *International glossary of hydrology: Geneva* (p. 385). Switzerland: World Meteorological Organization.

# Chapter 7
# Biodiversity

**Abstract** There are several aspects of biodiversity distributed worldwide. Some of them are widespread, while others are few due to the existing ecosystems. The Kingdom of Saudi Arabia, where diverse terrain properties exist, is characterized by defined climate and topography, and this in turn created diverse ecosystems influenced mainly by the arid climate, desert and mountainous, as well as the remarkable littoral elements. The existed biodiversity in NEOM Region can be described as "unique" because the ecosystems there are rarely found elsewhere and even the level of diversity is high, because the region contains abrupt topographic aspects and located dominantly between the desert and the sea. Nevertheless, the biodiversity and the existing ecosystems in NEOM Region have not studied in depth, but these must be well identified, notably that the region is being prepared to be a global hub. The highlight on the biodiversity, including all biological components in NEOM Region, will be a supportive tool for the SLM. Therefore, the smart orientation of biodiversity will attract people, and then it will add a significant value to the better use of the ecosystem in the region. This chapter will highlight the main components of biodiversity in NEOM Region. This was obtained depending on the available data sources as well as investigated elements of biodiversity aiming to reach preliminary picture on the subject matter.

**Keywords** Flora and fauna · Topographic changes · Desert · Arid area · Marine ecosystem

## 7.1 Introduction

Among Earth's biosphere, including lithosphere hydrosphere and atmosphere, there is a thick layer of approximately one kilometer where about 50 million species of animals, plants and micro-organisms live, and only 1.4 million of them were recognized. All these species are adapted to live in defined ecology which can be on mountain crests to deep marine water and from polar ice regions to deserts.

The biological diversity has been named as "biodiversity" by Walter Rosen in 1986 (UNEP 1995). Thus, biodiversity involves three major types: diversity within


M. M. Al Saud, *Sustainable Land Management for NEOM Region*,
https://doi.org/10.1007/978-3-030-57631-8_7

species (i.e. genetic diversity), diversity between species (i.e. species diversity) and diversity between ecosystems (i.e. ecosystem diversity).

### 7.1.1 Concepts

The diversity in living organisms and all biological elements is a normal life condition that always exists, but it differs by regions and at different levels, and this diversity is changeable in space and time, as well as the influencers, derivers and sources are tremendous. Therefore, biodiversity is everywhere in territorial and marine ecological systems, at different scales and speed of change. Thus, the virtual omnipresence of life on Earth is seldom appreciated because the largest number of organisms are too small (<5 cm) where they are sparse, ephemeral, or cryptic, and invisible to the unaided human eye.

There are many definitions for biodiversity, which is a term lunched in 1980. Thus, Swingland (2001) in Encyclopedia of Biodiversity defined biodiversity as "an attribute of an area; certainly refers to the variety within and among living organisms, assemblages of living organisms, biotic communities, and biotic processes, whether naturally occurring or modified by humans". Hence, it is represents the variability in the living organisms from whether the terrestrial or marine environment, and this includes variability within species, between species, and of ecosystems.

Biodiversity represents the basis of ecosystems by introducing a number of services that affect mankind life. It may contribute directly by living, regulating, and entertainment services or indirectly by the naturally induced services, as follows:

- Vital living services such as water, food, fiber and timber.
- Regulated services such as the adjustment of climatic condition, flood regime, disease spread and disappear, waste disposal, and regulating water quality.
- Entertainment services such as recreation, and promenade and enjoying nature.
- Naturally-induced services such as the formation of soil material, photosynthesis process, and nutrient cycling.

Therefore, the value of biodiversity is represented mainly by serving constituents of human well-being, including security, basic material for a good life, health, good social relations, and freedom of choice and action (Sarukhán and Whyte 2005).

The obtained studies on the changes in the economic value related to changes in biodiversity, such as forest fire, soil pollution, draining of wetlands, and wood log; concluded that the total economic cost of ecosystem conversion is significant and to sometimes it exceeds the benefits of the habitat conversion.

### 7.1.2 *Influencers on Biodiversity*

The change in biodiversity and ecosystem services is controlled by indirect and direct influencers. Thus, five indirect influencers can be illustrated and they are handled by human. These are:

1. Demographic: The increased population rise the consumption per capita and this increases loads and pressure on the ecosystems and biodiversity.
2. Economic: The system of trade and industry across the world (global economic activity) increased about seven times between 1950 and 2000, and expected to increase 4 times in 2050 (INEGI 2020). This will cause destruction and deterioration in many aspects of biodiversity.
3. Socio-political: The changing in governmental regulations create in consistent forms of management, notably the adaptive management, approaches.
4. Cultural: The cultural aspects play a major role in the ethical behavior towards natural resources and thus the biodiversity.
5. Scientific and technological: knowledge mapping and dissemination increase the efficiency in resource use; besides scientific knowledge provides the tool to increase exploitation of resources.

There are direct influences of biodiversity loss and change in ecosystem services. They ate governed by the region where they occur (Fig. 7.1). The most important direct influencers are:

1. Habitat change: This includes the changes in land use, changes in water bodies and green cover, as well as loss of coral reefs, and damage to sea floors.
2. Climate change: It involves the remarkable changes an oscillations in the climatic conditions and its impact on different ecosystems.
3. Invasive alien species: These are plants or animals that are introduced by man, accidentally or intentionally, outside of their natural geographic range into an area where they are not naturally present.
4. Overexploitation of species: It refers to harvesting a renewable resource to the point of decreasing returns. Thus, overexploitation may result resource destruction, including extinctions.
5. Pollution: It has many aspects (e.g. poison soils) that harmful to plants, animals and human as well. For example, the toxic chemicals that accumulate in top predators make some species unsafe to eat.

## 7.2 Factors on Biodiversity in NEOM Region

As mentioned previously, no studies have been done on biodiversity and ecosystem specifically for NEOM Region, except studies done on Tabuk Province/or the surrounding. It is utmost important to highlight on the existing biodiversity and ecosystems in this region as an additional component for adopting it as an area with

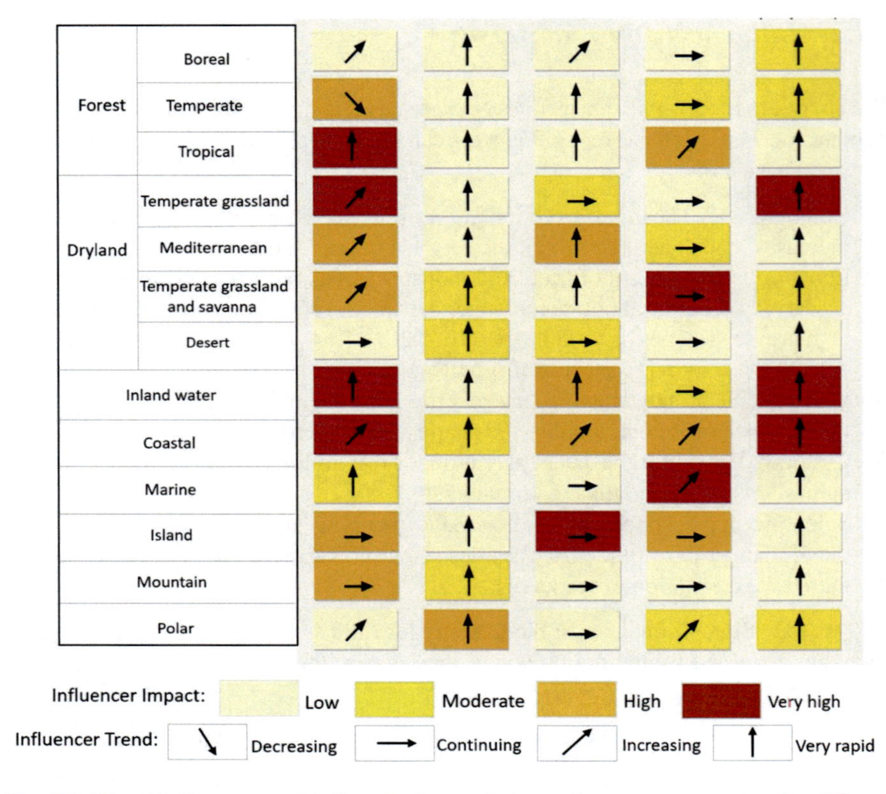


**Fig. 7.1** Direct influencers on biodiversity loss and change in ecosystem services in different regions (Millennium Ecosystem Assessment 2005)

remarkable ecologies. This would make it an attracting environment for people from different parts of the world.

NEOM Region is almost a bare area where water and wet soil are scarce, and this has been reflected on the little abundance of ecosystem, notably the flora. However, there are many factors controlling the existence and changing regime of biodiversity and the existing ecosystems (terrestrial and marine). These factors act at different levels of impact and even the time of influencing is also uneven. In addition, some of these factors may play an opposite role on biodiversity; for example, warm weather is harmful for some plant species, but it is also favourable for others and so on.

There are five main factors, in NEOM Region, controlling the biodiversity and ecosystems. These are:

### 7.2.1 Habitat

This represents the natural environment for many species where there species and organisms where they assure the presence of food, protection and mates for reproduction. All these habitats represent biogeographic units that control the distribution of life regime of different species, notably the flora and fauna. Hence, uneven biogeographic units create different species, like the case of NEOM Region.

For NEOM Region, there are diverse habitats occurred largely within the identified geomorphologic units (i.e. plateau, mountain ridges, coastal plain, and valleys). They exist as large-scale habitats including: mountains, desert, dunes, wadis, rocky outcrops, salt pans (sabkhas), coastal plain, tidal flats, littoral zone and many other habitats. Also, these are small-scale habitats which are included among the large-scale ones, such as: crests, gullies, caves, rock pockets, wetlands, soil intrusions, beach gravel, marine terraces, sandy sea floor, etc., which are dominant habitats in NEOM Region.

### 7.2.2 Climate

The climate of NEOM Region is distinguished by the topographic setting of the region where coastal plains extend at the foot slopes of mountain ridges in which the latter gradationally tilted eastward. This in turn creates a diverse climatic zone over different topographic units. Hence, the average rainfall rate is about 57.5 mm/year, but it ranges between 11 and 412 mm, and the average temperature is 22.5 °C, where the maximum reaches 44 °C and minimum 9 °C; in addition the area is known by high humidity since it is bordered from west and south by the sea.

Therefore, the distribution of biodiversity of animal and plant species is totally controlled by these topographic-induced climatic differentiation, and many of plant species for example are found with specific localities, but totally disappear in others.

### 7.2.3 Surface Materials

NEOM Region is characterized by intensive structures and tectonic movements which are reflected on the diverse lithological characteristics and distribution; however, surface materials (rocks and soil) are found with tremendous types and aspects. These materials play a significant role in the distribution of plant species (e.g. soil-related flora), notably the region contains bare rock (e.g. consolidated, friable, etc.), sand dunes, alluviums and detrital deposits, salty soil.

In addition, these materials make the region with active surficial processes, such as erosion and mass movement, which are also controlling the biodiversity and ecosystems. Figure 7.2 shows an example of active surficial processes in NEOM Region.

**Fig. 7.2** Example of active surficial processes in NEOM Region where sand waves of different types (evidenced by their colour) exist

## *7.2.4 Characteristics of Littoral Zone*

The characteristics of the littoral zone has several aspects including mainly the shoreline characteristics and materials, as well as seawater properties. In this respect, shoreline morphology (e.g. straight, irregular, curved, bay, etc.) is important as well as water depth and the located materials (beach sand, rocky each, gravel, etc.). In addition, the chemical and physical properties (e.g. salinity, hardness, density, DO, etc.) of seawater have significant role in the distribution and types of marine species. In addition, the physical processes occur in the sea are essential components in the aspect biodiversity and the marine ecosystems, and this includes, for example, sea currents, upwelling, sedimentation, sediments transportation, etc.

## *7.2.5 Human Interference*

Even though human is a part of biodiversity, but often he acts on the ecosystems in different ways. Thus human influences the biodiversity at different dimensions. In this respect, the anthropogenic interference is always considered whether for optimal or unfavourable ones.

Other than the necessary involvement, there is no remarkable human interference observed in NEOM Region. Thus, except the coastal zone, there is very few number of human settlements in the largest part of NEOM Region, and even in the coastal zone the existing ones are still few. Hence, the impact of human on the ecosystem of NEOM Region can be considered as negligible.

## 7.3 Geography of Terrestrial Biodiversity

As described in different studies, Tabuk Province (including NEOM Region) has a variety of plant and animal species forming a typical biodiversity and unique ecosystems. Their abundance and distribution are still undefined, notably in the rugged and remote areas due to the inaccessibility to reach these areas.

The utility of these species is still limited and can be considered as undiscovered. This is due to the reason that the region still is characterized by very low population density, and thus limited human activities, and no concerns are given to the region so far.

The inventory of biodiversity in NEOM Region, as illustrated in this document, was carried out depending on the available references as well as on field observations. Therefore, it can be considered as a preliminary assessment for the types and distribution of plant and animal species in this region, and yet in-depth surveys are required.

### *7.3.1 Plants Species*

The distribution of plants is mainly controlled by their responses to variation in environmental factors, including water availability, topography and soil; whereas some studies showed that topography and climate are the major factors affecting the degree of speciation (Abdel Khalik and El-Sheikh 2013; Osman et al. 2014).

The quantitative analyses of plant species extending from Tabuk Region to Red Sea coastal region and the Gulf of Aaqba was investigated by different researchers. Thus, Moawed and Ansari (2015) estimated a total of 82 plant species belong to 66 genera and 30 families were recorded. Besides, Al-Mutairi1 et al. (2016) reported a total of 96 species belong to 75 genera and 38 families (34 dicots and 4 monocots). While, Basahi (2018) counted 106 species for 86 genera and 36 family only along the Gulf of Aaqba. Therefore, compiling all identified families of plant species in NEOM Region results in 51 families. Hence, the families with more than 3 plant species are:

- Amaranthanceae: 24 species
- Poaceae: 18 species (example in Fig. 7.3)
- Asteraceae: 9 species
- Fabaceae: 8 species (example in Fig. 7.4)
- Brassicaceae and Zygophyllaceae: 7 species
- Chenopodiaceae and Poacea: 6 species
- Resedaceae: 5 species

**Fig. 7.3** *"Panicum turgidum"* belongs to Poaceae family. A dwarf shrub dominant on sand dunes and characterizes by Chamaephytic life form

**Fig. 7.4** *"Acacia tortilis"* belongs to Fabaceae family. A shrub tree dominant in wadis and characterizes by phanerophytic life form

- Lamiaceae: 4 species
- Boraginaceae: 4 species
- Asclepediaceae and Caryophyllacea: 3 species.

There are also 38 families with only one or two species. These are as follows:

- Acanthaceae
- Aizoaceae
- Apiaceae
- Asclep ediaceae
- Asphodelaceae
- Avicenniaceae
- Capparaceae
- Cistaceae
- Convolvulaceae
- Cucurbitaceae
- Euphorbiaceae
- Geraniaceae
- Grassulaceae
- Hyacinthaceae
- Liliaceae
- Mimosaceae
- Orobanchaceae
- Papaveraceae ← Identified only along the Red Sea
- Peganceae
- Plantaginaceae
- Salvadoraceae
- Scrophulariaceae
- Solanaceae
- Tamaricaceae
- Urticaceae
- Zygophyllaceae
- Cyperaceae
- Juncaceae
- Loranthaceae
- Malvaceae
- Menispermaceae
- Neuradaceae
- Nitrariaceae
- Nyctaginaceae
- Polygonaceae
- Portulacaceae
- Rhumnaceae
- Xanthorrhoeaceae.

(Cyperaceae to Xanthorrhoeaceae: Identified only along the coast of the Gulf of Aaqba)

The applied chorological analysis of the recognized species indicated the predominance of mono-regional taxa over the other elements. Moreover, the chorology of flora of Tabuk Region showed that most species belonged to Saharo-Arabian (37.21%), Irano-Turanian (11.63%) and Sudanean (10.47%) elements, and this comprise almost 60% of the total number of plant reported species (Al-Mutairi et al. 2016).

The Chamaephyte and Theorphytes taxa constitute the dominant life forms with approximately 69%; followed by Geophytes, Hemicryptophytes, Phanerophytes and Parasites taxa with 13, 10% and 8%; respectively (Al-Mutairi et al. (2016). While, the life form along the coastal zone of the Gulf of Aaqba is represented by Halophytes in the coastal flats, Ephemerals in Wadi beds and Chamaephyte along foothills and slopes (Basahi (2018).

The highest number of species (i.e. 57.32%) was recorded for perennial, while annual species was recorded 42.68% (Moawed and Ansari 2015).

There are several plant species which are well pronounced in the region for their traditional and medical use to treat the digestive tract diseases and parasites. They are characterized by the uniqueness in diversity of habitats such as mountains, sand dune, wadis and coastal flats. They mainly belong to 30 families including, Asteraceae (Example in Fig. 7.5), Fabaceae, Aizoaceae and Zygophyllaceae (Alharbi 2017).

**Fig. 7.5** *"Anvillea garcinii"* belongs to Asteraceae family. A medical shrub used their leaves and seeds to treat gastro and intestinal troubles

### 7.3.2 Animal Species

In the Kingdom of Saudi Arabia fauna has given much attention than flora. This is due to interest in the big mammals for the purpose of hunting and shooting, as well as the husbandry of animals as old customs and traditions. In addition, birds and butterflies have been also investigate, but less is known about other animals.

Likewise the case of plant species, there are few studies done on animal species in Tabuk Province including NEOM Region. However, the study of animal spices is more difficult than that for plants due to the reason that the mobility of animals needs regulated monitoring using tracing, fixed cameras and traps as well, and lately drones become very useful tool in this respect. Thus, to inventory for the existing animal species; physical description, biodiversity index and the relative abundance and species richness must be elaborated.

1. Major animal groups

According to Badawi (2012), Tabuk Region includes the following 7 major animal groups. Hence, the biotic factors on animal communities were measured by calculating the diversity index (D*i*) which indicates how many different types exist, and thus it based on counting the total number of individual (N) and the number of individual of spices (n). Therefore, D*i* is represented by the following formula:

$$Di = N(N - 1)/\Sigma n(n - 1)$$

Therefore, the following spices, with their D*i* and were accounted:

- Birds: 72 species (D*i* =3.05)
- Mammals: 10 species (D*i* =1. 5)
- Reptiles: 18 species (D*i* =2.3)
- Amphibians: 7 species (D*i* =1.4)
- Invertebrates: 13 species (D*i* =2.23)
- Insects: 93 species (D*i* =3.86).

Besides, another study reported that in Tabuk Region, there are 35 mammal species, 37 reptile species and 167 bird species constituting 82 resident, and 85 migratory birds have been recorded from Tabuk region (Balletto et al. 1985; Arnold 1986; Gasperetti 1993; NCWCD 2000).

In addition, the relative abundance of species was also calculated by dividing the number of individuals in one type over the total number of individuals in all species. Hence, an inverse relationship was found between species diversity and the relative abundance as evidenced by Badawi (2012) who also came to the following findings:

- Birds: The abundance was evidenced by the dominant species of pigeon and sparrows, while other species are rare.
- Mammals: The higher abundance was found in Gerbillus nanus while the least in Spalax leucodon.

- Reptiles: Stenodactylus arabicus and stenodactylus grandiceps are the most abundant species.
- Amphibians were the highest percentage of 21%.
- Invertebrates: There is convergent abundance of spiders except in Pholcus phalangioide and also in Scorpions in Leiurus quinquestraitus and Compsobuthus arabicus.
- Insects: There are tremendous species over the whole year, obvious abundance in summer and least in winter.

2. Large mammals

Usually large mammals are given attention due to their obvious existence, and sometimes due to their benefit use for human. In Saudi Arabia, large mammals are distinguishable due to their biodiversity and wide spread in different ecologies.

For NEOM Region, where studies on large mammals are rare, the existing mammals can be divided into Herbivorous and Carnivores.

- Herbivorous

Except dromedary (camels), Herbivorous are rarely found in NEOM Region, but sometime they exist with very little abundance. If gazelles are considered, this can be attributed to little the few number of studies done in the region, and their presence in remote areas.

- Camelidae: This represents dromedary where the total population of dromedary is estimated to be around 1.6 million camels within the Arabian Peninsula where approximately 53% found in Saudi Arabia (FAO 2000). Thu, the Arabian camels (Camelus dromedaries) are widespread in Saudi Arabia.
  Based on the breed classification adopted by Abdalla and Faye (2012) which considers local naming, the commonly found Mustellidae: the ratel Mellivora capensis (examples in Fig. 7.6).
- Bovidae: This includes antelopes, cattle, gazelles, goats, and sheep. In Saudi Arabia, there are 140 species of wild and domesticated animals in this family. For cattle, goats and sheep, they likely are not largely widespread in the study area. Beside, Arabian and Sand Gazelles and Arabian were reported. Thus, Dorcas gazelles, i.e. Al Afri; Capra ibex, i.e. Wail and mountain Gazelle, i.e. Al Admi (Al Nafie 1982).
- Eauidae: This animal family belongs to horses and donkeys, which are rarely found in the area of study, except some Hassawi donkey (i.e. generally white color) and Baladi donkey (i.e. grey and smaller than the former one).

- Carnivores

Carnivores in Tabuk Province were identified in a comprehensive study carried out by Aloufi and Amer (2019), based on camera traps, live traps and direct observations. Hence, 7 species of carnivores representing four families were identified. These are:

**Fig. 7.6** Dromedaries (Waddah, white and Sahael, brown) common breeds in NEOM Region

- Felidae: This includes (1) Felis Margarita which is located in sand dune and the adjacent rock caves and pockets, (2) Panthera pardus, which is a leopard reported in the coastal zone of Tabuk (Al-Johany 2007).
- Hyaenidae: The common type found in the study area is Hyaena which is a striped hyena.
- Mustellidae: the ratel Mellivora capensis was reported in Tabuk Province for the first time in 1985 (Gasparetti et al. 1985).
- Canidae: Despite being a regionally endangered animal species, wolf is considered as one of the most persecuted animals by locals. The identified species are: (1) Canis lupus, (2) Vulpes Cana and (3) Vulpes, red fox.

## 7.4 Geography of Marine Biodiversity

There are some studies done on the marine ecosystems for the Gulf of Aaqba and the Reds Sea where NEOM Region occupies the largest part at the Gulf of Aaqba and relatively small part along the Red Sea.

- The Gulf of Aaqba marine ecosystem

There is about 115 km coastline of NEOM Region along the Gulf of Aaqba where the marine ecosystem is characterized by rich and diverse habitats with over 1000 species of tropical fish, 110 species of reef building corals and 120 species of soft corals (Ammar et al. 2013).

1. Coral reefs: There are 158 coral species belong to 51 genera and 12 families. Hence, the studied sites along entire the Gulf of Aaqba by Ammar et al. (2013), resulted in identifying coral reefs belong to the following types:

   - Stony corals (e.g. Acropora granulosa, Ctenactis echinata, Goniastrea retiformis, Mycedium elephantotus, Polcillopor damicornis, Acropora tenuis, example Fig. 7.7).
   - Soft corals (e.g. Heteroxenia ghardaqensis, Lobophytum, example Fig. 7.7).
   - Hydrocorals (e.g., Millpora platyphylla, Millepora dichotoma, example Fig. 7.7).
   - Gorgonians (e.g. Anella gorgonian, Anella glauca, Xenia umbellate, example Fig. 7.7).
   - Black coral (e.g. Antipathies, Protoptilum).

2. Fishes: According to Khalaf (2004), the total fish species along the Gulf of Aaqba is 507, and they belong to 109 families with an average of 4.7 species per family; 18 Chondrichthyes and 489 Ostichthyes. The largest families are as follows:

   - Labridae: 51 species
   - Pomacentridae: 29 species

**Fig. 7.7** Coral reefs from the Gulf of Aaqba. (a. Acropora tenuis), (b. Lobophytum), (c. Millepora dichotoma), (d. Xenia umbellate)

- Serranidae: 25 species
- Apogonidae: 24 species
- Blenniidae: 24 species
- Gobiidae: 21 species
- Carangidae: 17 species
- Syngnathidae: 16 species.

Detailed inventory on fish families and their species were reported by Fishbase (2019). Examples of some families such as: Synodontidae, Acanthuridae, Sparidae, Alopiidae, Mullidae, Belnniidae, Lethrinidae, Uranoscopdiae, Xenisthmidae, Triglidae and Platycephalidae.

3. Mangroves: Few studies mentioned the presence of mangrove along the Gulf of Aaqba (Kouchzius 2002). Whereas, on the facing coast of Sinai, some studies illustrated quantitative classifications for mangroves (example: Abubkr 2012)
4. Seagrass: Only 8 species of seagrass have been reported along the Gulf of Aaqba (El Shaffai 2011). These species were grouped into two Cymodoceaceae and Hydrocharitaceae.

- Red Sea marine ecosystem

The coast of NEOM Region along the Red Sea is about 110 km. The Red Sea is almost a semi-enclosed and elongated water body of relatively warm water, and this governs the existing marine ecosystem.

1. Coral reefs: According to De Vantier et al. (2000), there are approximately 260 coral species in 68 genera of 16 families in the Red Sea. The most common families can be summarized as follows:

   - Acroporidae: 64 species
   - Faviidae: 61 species
   - Poritidae: 27 species
   - Funggiida: 26 species
   - Agariciidae: 21 species
   - Mussidae: 15 species
   - Pocilloporidae: 11 species
   - Siderastreidae: 10 species
   - Dendrophyllidae: 10 species
   - Pectinidae: 7 species
   - Merulinindae: 3 species
   - Asrtocoeniidae: 2 species
   - Oculinidae: 2 species.

2. Fishes: About 1120 fish species belong to 159 families exist in the Red Sea within an overall depth ranges between 0 and 200. Thus, 165 species of them are exclusively endemics to the Red Sea (Najeeb et al. 2018).
   The following are the most dominant fish families and their species:

- Gobiidae: 24 species
- Blenniidae: 20 species
- Leiognathidae: 12 species
- Syngnathidae: 11 species
- Callionymidae: 10 species
- Scombridae: 10 species
- Apoginidae: 9 species
- Cynoglossidae: 9 species
- Nemipteridae: 9 species
- Micodesmedae: 8 species
- Exocoetidae: 8 species
- Bothediae: 6 species
- Sparida: 6 species.

3. Mangroves: In the c mangroves. This is due to the high salinity, poor soil textures, low precipitation and nutrient concentrations. They are represented by three species: Avicennia marina, Rhizophoramucronata, and Bruguieragymnorhiza (Mandura 1997).
4. Seagrass: According to Barale Vittorrio (2007), ten species of seagrass are widespread in the Red Sea. There are more than 70 species of seagrasses and they are classified into two families: Cymodoceaceae and Hydrocharitaceae (Serrano et al. 2018).

## 7.5 Ecosystem Loss in NEOM Region

Studies on the loss and deterioration of ecosystems are found with considerable number if compared with studies on the general assessment and quantitative analysis of biodiversity in Tabuk Province and the surrounding. This can be, sometimes, normal because the support of biodiversity to ecosystem services is usually faced by natural and man-made constraints. Hence, the biodiversity with its all ecological branches always witnesses loss in its components.

NEOM Region, can be considered as an area with few features of biodiversity loss. This can be attributed to the fact that the urban settlements are this region is sparse. However, to highlight on the loss of ecosystems at its initial stage would be significant and helpful for future precautionary implements, notably NEOM Region has assigned as a global site and vulnerable ecosystem should be identified.

The loss the ecosystems in NEOM Region exists at different levels in space and time and it differs between the located species. In a general assessment, loss of ecosystems is well pronounced in animal species, notably the Herbivorous and Carnivores, as a result of human behavior. In addition, the marine ecosystem is the second impacted biodiversity in NEOM Region while the plant spices in the least influenced.

The features and info about the loss of ecosystems in NEOM Region can be summarized as follows:

**Fig. 7.8** Chaotic construction and encroachment along the southern coast of NEOM Region

1. Damage in the vegetation cover in many localities due to off-road driving and sometimes due to overgrazing.
2. Unmanaged and chaotic construction and urban encroachment towards the coast (example Fig. 7.8).
3. There are several aspects of marine pollution, notably the oil spills from ships, as well as the sludge.
4. Illegal fishing and over fishing in some localities, in addition to the use of alien species.
5. Illegal animal trading is known in Tabuk Region where, for example, Peregrine Falcon costs $ 7300, Gyrfalcon $ 3070, Arabain wolf $ 800 (Aloufi and Eid 2014).
6. Animals hunting is pronounced in the region and this acts on the extinction of many species.
7. Distraction of coral reefs and other shallow-water habitats through inadequate anchoring practices.
8. Impact of climatic extremes on the ecosystems is well pronounced (e.g. erosion, landslides, desertification, etc.).

## 7.6 Natural Reserves in NEOM Region

NEOM Region has several potential natural resources including the geologically-related ones (as discussed in Chap. 6); in addition to the resources with ecosystem advantages, such as the natural reserves where unique flora and fauna exist as well as the distinguished landscape and diverse natural components.

As concluded in previously, the existing ecosystems with their biodiversity in NEOM Region are unique of their types, and they are under slow processes of biodiversity loss; however, unless these ecosystems are conserved and the process of loss (deterioration, destruction, extinction) is stopped, otherwise they will reduce their uniqueness. Therefore, the protection of these ecosystems became a necessity to conserve the natural components that this region occupies.

### *7.6.1 General Overview*

In fact, the Kingdom of Saudi Arabia encompasses many diverse natural environments and sites where each of them can be a landmark. Hence, there are several areas which encompass distinguished ecological components. They have been designated as protected areas (i.e. natural reserves); and the concept of natural reserves has been adopted since 1987 when Harrat Al Harrah was considered as a natural reserve. Therefore 15 natural reserves were adopted since then. Thus, Saudi Arabia carried out the required environmental and social studies in 1991, as well as feasibility studies to prepare the system of protected areas with the experience of the International Union for Conservation of Nature (IUCN). This system, which has been recently updated according to environmental developments, includes the proposal to protect 75 areas where there are 62 terrestrial 13 coastal and marine areas.

Later on, the Saudi Wildfire Authority (SWA) designated 15 natural reserves (SWA 2020) where two of them are new areas (Jubail, Majami and Jabal Shada) other than reported by the National Commission for wildlife Conservation and Development (NCWCD). In addition, there are two natural reserves were excluded by SWA from to the list. These are Al Janadreyah and Al Hair wetland (Table 7.1).

The total area of these natural reserves (protected areas) is about 102361 km$^2$. Besides, the largest one is Al-Khunfah with 19339 km$^2$, and the smallest one is Jabal Shada with 69 km$^2$. While, the closest natural reserves to NEOM Region are Al-Khunfah and Al-Tubaya.

### *7.6.2 Components of Natural Reserves*

There are many ecological and environmental characteristics in NEOM Region which are almost similar of those in the designated natural reserves of Saudi Arabia. This

**Table 7.1** Natural reserves in the Kingdom of Saudi Arabia (SWA 2020)

| # | Protected area | Environment | Designation date | Area (km$^2$) |
|---|---|---|---|---|
| 1 | Jubail | Terrestrial | 2019 | 2411 |
| 2 | Al-Khunfah | | | 19339 |
| 3 | Ibex Reserve | | | 18409 |
| 4 | Mazra'at As-Sayd | | | 2553 |
| 5 | Majami Al Hadb | | | 2256 |
| 6 | Raydah | | | 9.5 |
| 7 | Harrat Al Harrah | | | 13775 |
| 8 | Jabal Shada | | 2001 | 69 |
| 9 | Saja Um Al-Rimth | | 1994 | 6528 |
| 10 | At—Taysiyah | | | 4272 |
| 11 | Nafud Al Urayq | | | 2036 |
| 12 | Uruq Bani Ma arid | | 1992 | 12787 |
| 13 | Al- Tubaya | | 1988 | 12105 |
| 14 | Farasan Islands | Marine | | 5408 |
| 15 | Umm al-Qamari | | 1987 | 403 |

motivated proposing new natural reserves in NEOM Region, notably that the natural resources, including unique ecosystem, in this region must be identified.

A nature reserve can be described as a wildlife refuge, wildlife sanctuary, protected area, biosphere reserve, biological reserve or nature conservation area. Thus, a natural reserve is created by the human initiative to protect an area with significant flora, fauna, or features of geological or other special interest. It is; therefore, reserved and managed for purposes of conservation.

The major natural components to assign a natural reserve can be summarized as follows:

1. Biological uniqueness

This includes the existence of unusual and rare aspects of biological life including distinguished species of fauna and flora. This is well pronounced in NEOM Region where a number of unique species were reported including 51 plants families with 106 species, whereas animals were found in six groups representing 213 species.

Some of the identified plant species are endemic and exist only in the Red sea, such as Papaveraceae, while others ones are endemic only for the Gulf of Aaqba, such as Cyperaceaea and Juncaceae. This is also the case for the animal species where unique Herbivorous and Carnivores species exist and some of them are endemic such as many types of Bovidae (e.g. Dorcas gazelles, Al Afri, Al Admi, etc.).

2. Necessity for protection

Likewise many areas in the Kingdom, there are many types of fauna and flora in NEOM Region threatened by extinction, and some others are already became extinct and hence there is necessity for protection.

Even though, no accurate measurable information about the extinct animal and plant species in Tabuk Region; however, a total of 9 species threatened with extinction were reported in the Red Lists of IUCN (2012), including Critically Endangered Reptilian, Endangered bird species, 5 Vulnerable species and 2 Near Threatened species.

3. Virgin lands

There are many lands with unique fauna and flora, but constraints exist to protect these lands. These constraints can be due to: proximity to urban activities, lands cut by major routes (railways, major roads, pipelines, etc.) and international shared lands. Hence, remote and virgin areas are usually most desirable to be natural reserves.

4. Protection ability

In many cases, naturally-characterized areas could not be reserved due to the lack to logistics, financial resources, and personnel for guard and the lack to creditable studies done to for the assignment of natural reserves. Unlike the case in Saudi Arabia where reserving these areas has been adopted and all requirements have been secured.

### *7.6.3 Proposed Natural Reserves*

Any existing resource will add a significant value to NEOM Region. Therefore, the presence of natural reserves in NOEM Region will give it additional advantage, and the proposed natural reserves will be a shrine for visitors, researcher and explorers.

In spite that many areas and localities can be designated as natural reserves in NEOM Region; however, the author proposed five natural reserves depending on their geographic location, the bareness of landscape, minimal human interference as well as on the field observations. Therefore, the concept behind proposing these natural reserves came from the necessity to utilize any existing natural resources including ecological resources. In this respect, in-depth surveys should be carried out, including types of species and their geographic distribution, to make the final decision about the designated natural reserve in this document.

The proposed five natural reserves in this document include three of terrestrial ones which are mainly located in the mountainous ridges and two marine ones for Thiran and Sanafeer Islands (Fig. 7.9). The selection of the geography of these reserves considered components illustrated in the previous section (Sect. 7.6.2). The

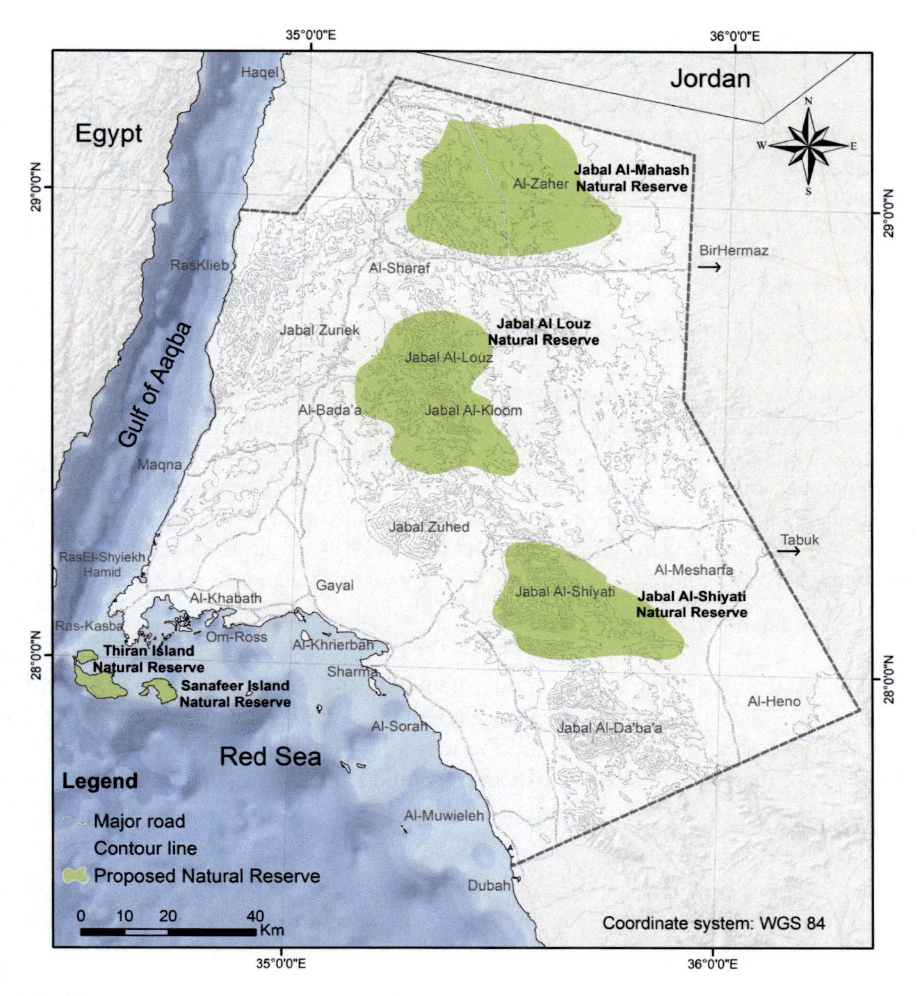


**Fig. 7.9** Propose natural reserves in NEOM Region

total area of the terrestrial ones is approximately 2950 km$^2$, while the marine reserves occupy 113 km$^2$.

1. Jabal Al-Mahash Natural Reserve

The surface area of the Jabal Al-Mahash Natural Reserve is about 1155 km$^2$. It is located in the most northern part of NEOM Region, along a number of mountains including Jabal Zaher (1387 m), Jabal Om Louza (1623 m), Jabal Al-Nomierh (1353 m) and Jabal Amiq 1766 m). The selected area is cut by a number of wadis, specifically Wadi Al-Abyad, Wadi Jnaf, Wadi Wassat and Wadi Al-Dahyel (Fig. 7.9). This area includes two main springs, Al-Nejieleh Spring and Al-Katar Spring.

2. Jabal Al-Louz Natural Reserve

Located in the middles part of NEOM Region, this natural reserves has an area of about 980 km$^2$ where mountain hills and crests are dominant. The bench make here is represented by Jabal Al-Louz (2401 m) and Jabal Fayhan (2549 m), the heights elevations in NEOM Region. In addition, there are other mountians exist such as Jabal Al-Kloom, Jabal Alhouief and Jabal Omm Hayfa. Also the area occupies several incident wadis span between the existed mountains and most of these wadis convert from this area, such as Wadi Moussa, Wadi Rayt and Wadi Al-Abyad (Fig. 7.9).

3. Jabal Al-Shiyati Natural Reserve

This is located in the southern part of NEOM Region with an area of about 815 km$^2$. It is located in the mountain ridges and include a part of the plateau to the east (Fig. 7.9). Hence, Jabal Al-Shiyati is the most elevated peak in this region with 2103 m, plus many other mountains such as Jabal Terban (1612 m), Jabal Omm Haytham (1524 m) and Jabal Nakhleh (1576 m). While many major wadis exist including Wadi Arnab, Wadi Al-Malas and Wadi Omm Arrarah. This area, with the presence of shallow groundwater have some springs, mainly Ain Nimah Spring and Bir Al-Zereb Spring.

4. Thiran Island Natural Reserve

Located on the gate of the Gulf of Aaqba, Thiran Island has an area of about 80 km$^2$ (Fig. 7.9). It has a unique marine ecosystem and surrounded by shallow seawater extends several kilometers to its northern side at Ras Qasba. Thus, the highest elevation of the island is 518 m exists at its southern side.

5. Sanafeer Island Natural Reserve

With a distance of about 2.5 km from Thiran Island, Sanafeer Island is located and has an area of about 33 km$^2$. The highest elevation in this island is 112 m, and it is totally surrounded by shallow marine water over few kilometers.

## References

Abdalla, H., & Faye, B. (2012). Phenotypic classification of Saudi Arabian camel (Camelus dromedarius) by their body measurements. *Emirian Journal of Food and Agriculture., 24*(3), 272–280.

Abdel Khalik, K., & El-Sheikh, M. (2013). El-Aidarous, A. *Turkish Journal of Botany., 37,* 894–907.

Abubkr, S. (2012). *Ecological Study of Mangrove Forests (Avicennia marina (Forssk.) Vierh.) In South Sinai, Egypt*. (Ph.D. dissertation). Philadelphia University, Jordan. 213 pp.

Al Nafie, A. (1982). *Large mammals of central and north Saudi Arabia (Biogeographic study).* (MSc Dissertation). The Islamic University of Imam Mohammed Ibn Saud. 206 pp.

Alharbi, N. (2017). Survey of Plant Species of Medical Importance to Treat Digestive Tract Diseases in Tabuk Region, Saudi Arabia. *Journal of King Abdulaziz University-Science, 29*(1), 51–61. https://doi.org/10.4197/Sci.29-1.6.

Al-Johany, A. (2007). Distribution and conservation of the Arabian leopard Panthera pardus nimrin Saudi Arabia. *Journal of Arid Environments, 68,* 20–30.

Al-Mutairi, K., Al-Shami, S., Khorshid, Z., & Moawed, M. (2016). Floristic diversity of Tabuk province, north Saudi Arabia. *The Journal of Animal and Plant Sciences, 26*(4), 1019–1025.

Aloufi, A. A., & Amr, Z. S. (2018). Carnivores of the Tabuk Province, Saudi Arabia (Carnivora: Canidae, Felidae, Hyaenidae, Mustelidae). *Lynx, new series, 49*(1), 77–90. https://doi.org/10.2478/lynx-2018-0010.

Aloufi, A., & Eid, E. (2014). Conservation Perspectives of illegal animal trade at Tabuk Local Market, Kingdome of Saudi Arabia. *TRAFFIC Bulletin, 26*(2), 77–80.

Ammar, M., Emara, A., Nasser, M., & Al-Azim, H. (2013). *Marine life and chemistry of the Gulf of Aqaba and Ras Mohammed (Published Book)* (p. 75). Germany: Lap Lambert publishing.

Arnold, E. (1986). A key and annotated checklist to the lizards and amphibians of Arabia. *Fauna of Saudi Arabia, 8,* 385–435.

Badawi, A. (2012). *Study of Animal diversity in Tabuk region, Saudi Arabia.* (MSc dissertation in biological sciences). King Abdulaziz University. Jeddah, KSA.

Balletto, E., Cherchi, M., & Gaspertti, J. (1985). Amphibians of the Arabian Peninsula. *Fauna of Saudi Arabia, 7,* 318–392.

Barale, V. (2007). Marine and coastal features of the Red Sea. *European Commission, EUR 23091,* 56.

Basahi, R. (2018). Plant diversity of the coastal regions of Gulf of Aqaba, Saudi Arabia. *Annual Research and Review in Biology, 26*(3), 1–11.

De Vantier, L., Turak, E., Al-Shaikh, K., & De ath, G. (2000). Coral communities of the central-northern Sausi Arabian Red Sea. *Fauna of Arabia, 18,* 23–66.

El Shaffai, A. (2011). *Field guide for seagrasses of the Red Sea.* Courbevoie, France: IUCN, Gland, Switzerland and Total Foundation.

FAO. (2000). *2000 World Census of Agriculture. Main Results and Metadata by Country.* 246 pp.

Fishbase. (2019). *Fish Species in Gulf of Aqaba.* Available at: https://www.fishbase.se/search.php.

Gasparetti, J., Harrisond, L., & Büttiker, W. (1985). The carnivores of Arabia. *Fauna of Saudi Arabia, 7,* 397–461.

Gasperetti, J. (1993). Snakes of Arabia. *Fauna of Saudi Arabia., 9,* 169–450.

INEGI (The Instituto Nacional de Estadística y Geografía). (2020). Global Economic Activity Indicator. Avaialable at: https://en.www.inegi.org.mx/temas/igae/.

IUCN (International Union for Conservation of Nature). (2012). IUCN Red List of Threatened Species. Available at: https://www.iucnredlist.org/.

Khalaf, M. (2004). Fish Fauna of the Jordanian Coast, Gulf of Aqaba, Red Sea. *King Abdulaziz University: Marine Science, 15,* 23–50.

Kouchzius, M. (2002). Coral reefs in the Gulf of Aqaba. In C. Wilkinson (Ed.), *Status of coral reefs of the world: 2002.* Australian Institute of Marine Science.

Mandura, A. A. (1997). A mangrove stand under sewage pollution stress: Red Sea. *Mangroves and Salt marshes, 1*(4), 255–262. https://doi.org/10.1023/A:1009927605517.

Millennium Ecosystem Assessment. (2005). *Ecosystem and human well-being. Biodiversity synthesis.* A Report of the millennium ecosystem assessment board. World Resources Institute. Available at: http://www.millenniumassessment.org/documents/document.354.aspx.pdf.

Moawed, M., & Ansari, A. (2015). Wild plants diversity of Red Sea coastal region, Tabuk, Saudi Arabia. *Journal of Chemical and Pharmaceutical Research, 7*(10), 220–227.

Najeeb, R., Stewart, I., Vine, P., & Nawab, Z. (2019). Introduction to oceanographic and biological aspects of the Red Sea. In *Oceanographic and biological aspects of the Red Sea.* Cham: Springer.

NCWCD (National Commission for Wildlife Conservation and Development). (2000). *The study on coastal/marine habitat and biological inventories in the northern part of the Red Sea coast in Saudi Arabia.* Final report. Unpublished.

Osman, A., Al-Ghamdi, F., & Bawadekji, A. (2014). Saudi Journal of Biological. *Science, 21,* 554–565.

Sarukhán, J., Whyte, A. (2005). *Ecosystem and human well-being. Biodiversity synthesis.* A Report of the millennium ecosystem assessment board. World Resources Institute, Washington, DC. 86 pp.

Serrano, O., Almahasheer, H., Duarte, C. M., & Irigoien, X. (2018). Carbon stocks and accumulation rates in Red Sea seagrass meadows. *Scientific Reports, 8,* 15037. https://doi.org/10.1038/s41598-018-33182-8.

SWA (Saudi Wildfire Authority). (2020). Protected Areas. Available at: https://swa.gov.sa/En/Wildlife/ProtectedAreas/Pages/default.aspx.

Swingland, I. (2001). Biodiversity, Definition of. *Encyclopedia of Biodiversity*, 377–391. https://doi.org/10.1016/b0-12-226865-2/00027-4.

UNEP. (1995). *Global biodiversity assessment*. Cambridge University Press.

# Chapter 8
# Natural Hazards

**Abstract** Natural hazards have become the foremost geo-environmental issue in several regions on the Globe. Hence, it is rarely a month goes by without hearing about a disastrous event harming both urban structure and human life as well. The impact of natural hazards is being increased by the influence of many physical and man-made factors. There are several aspects of natural hazards, where some of them occur as flash events, other with relatively long time and other ones take longtime to feel its impact. The identification of natural hazards becomes an essential component to be considered while studying the sustainable land management programs. This is right enough, because the development of new urban and commercial plans should consider the natural security and in many cases, the geography of new designated urbanism is changed when it becomes endangered by physical processes, or there can be engineering controls applied to reduce and mitigate the impact of these hazards. NEOM Region, gives a typical example for the consideration of natural hazards in the SLM, notably this region is presumed to be an holding hub for different aspects of human activities. There are three major threatening hazards in NEOM Region, including mainly: seismic activities, floods, and terrain instability. This chapter introduces a comprehensive assessment and mapping for these three hazards which should be a perquisite for further SLM approaches. The generated data and information in this chapter were elaborated using time series records and measurements, plus space techniques and geo-information systems.

**Keywords** Smart city · Torrential rain · Earthquake · Urban planning · Commercial hub

## 8.1 Introduction

The planning processes in area under development do not usually account measures to mitigate natural hazards, and thus the resulted consequences, caused by natural disasters, are needless human suffering and economic losses. From the early work phases, planners should investigate natural hazards as they prepare investment projects and


M. M. Al Saud, *Sustainable Land Management for NEOM Region*,
https://doi.org/10.1007/978-3-030-57631-8_8

should promote approaches of avoiding damage caused by any anticipated natural risk.

Nature is typified by continuous changes and fluctuations, which are sometime by predictable evolution or the normal sequence of cyclical events as in seasonal weather. When unpredictable natural event becomes extreme in it occurrence, it therefore constitutes a danger to human and to the other members of the environment, such event is then described as a natural hazard (Al Saud 2018a).

"Natural Hazards" often used as a head title for several topics of researches and projects including the forecasting of disastrous events, risk management, and the technological controls to face these unfavorable physical processes. Foremost, the definitions and types of natural hazards must be made clear.

### *8.1.1 Definitions and Concepts*

Natural hazards occupy significant challenge in many regions, especially where there is high frequency of recurrence of these hazards and where the impact is harmful enough to cause the communities under lurking risk. Another way of conceptualizing natural hazards is the coexistence of humans in a normal environment that may threaten their safety, property at any time. Thus, the concept "natural hazard" is brought to mind when the word "disaster" is mentioned (Al Saud 2018a).

There are many definitions describing natural hazards; for example Burton et al. (1978) defined them as: those elements of the physical environment, harmful to man and caused by forces extraneous to him. While, the Asian Disaster Preparedness Center (ADPC 2000) described a natural hazard as a threat, and a future source of danger. Thus, it has the potential to cause harm to:

- People—decease, injury, disease, pandemic and stress
- Human activity—trading, economic, life style, educational etc.
- Property—property damage, economic loss of
- Environment—loss fauna and flora, pollution, loss of comforts.

Naturally occurred hazards may originate in different sources and physical systems, such as atmospheric, hydrologic, oceanographic and tectonic systems whereas the damaging impacts are equally catastrophic in many cases. This motivate the necessity for the interaction between different scientific themes and operational disciplines, eventually aimed at enhancing the reduction and mitigation of hazards.

There are several concepts known on natural hazards or natural disasters, but all of them describe it as the risk caused by nature's physical processes. These processes usually create damages to the environment and in many instances, they may result in hundreds of deaths and cost several of billions of dollars, disruption of commerce, and destruction of homes and critical infrastructure (Al Saud 2018a).

Most impacts caused by physical processes are in relationship to each other. For example, intensive erosion processes are almost caused by floods and torrents. Besides, some natural hazards has one aspect of impact and damage, such as that drought which is linked only to famine, but the case of floods and earthquakes are totally different. Also, natural hazards differ between regions, and they are largely induced by the existing natural influencers and partially by the anthropogenic ones; and thus, the magnitude of the impact is also simultaneously diverse. In this regard, there are regions with multiple hazards when more than one hazard event impacts the same area.

The "Return period" is always considered and calculated in the management plans. Hence, the retune period has a majority on human time-scale. For example, there are seven-year flood, or decade earthquake. This reflected by the statistical records of how often a hazard event of a define intensity will occur. Thus, the recurrence of hazard is a functioned by its frequency.

### *8.1.2 Types of Natural Hazards*

Different types of natural hazards occur with different mechanisms and different impacts. However, the classification of natural hazards types often contradictory and follows diverse aspects. Some of them accounted for the regime and origin of the physical processes, others consider the recorded magnitude and degree of damages, and so on. In regard to the types of natural hazards, a typology has been elaborated by Hewitt and Burton (1971) as follows:

1. Atmospheric:
    - Single element
        - Excess rain
        - Freezing rain (glaze)
        - Hail
        - Heavy snowfall
        - High wind speed
    - Combined elements
        - Hurricanes
        - Glaze storms
        - Thunderstorms
        - Blizzards
        - Tornadoes
        - Heat/cold stress
2. Hydrologic
    - Flood- river and coastal
    - Wave action
    - Drought
    - Rapid glacier advance

3. Geologic:
    - Mass movement (e.g. land slide, mud flow, etc.)
    - Earthquake
    - Volcanic eruption
    - Rapid sediments movement
4. Hydrologic:
    - Epidemic in humans
    - Epidemic in plants
    - Epidemic in animals
    - Locusts
5. Technologic:
    - Transport accidents
    - Industrial explosions
    - Fire accidents
    - Nuclear accidents
    - Collapse in buildings

Humans can do little or nothing to change the incidence of many natural processes, but they have a significant role to ensure that these processes are not converted into disasters. Hence, human intervention remains an influencer on natural hazards. This intervention can: (1) increase the frequency and severity (2) cause natural hazards where none existed before and (3) reduces the mitigating effect of natural ecosystems (ADPC 2000).

The most significant aspects of natural hazards are: floods, mass movement, earthquakes, volcanoes, and many others but with either less impact or recurrence (e.g. tornadoes, dust storms, tsunamis). Some of these hazards exist very slowly and then described as "creeping hazards, such as desertification and drought, while others occur extremely fast where they described as "flash events", such as earthquakes, volcanoes. Other hazards take place in limited intervals of time, such as flood, erosion and some aspects of mass movements.

In regard to the regional setting, the sources of vulnerability to natural hazards in most countries of the MENA Region (Middle East and North Africa) include mainly water scarcity, drought climate variability and change, as well as the increased human activities, specifically that the urban population already accounts for 62% of the total population, and this is anticipated to double by 2040. Additionally, 3% of the region's area is home to 92% of the total population. Therefore, people in the urbanized areas should be always prepared for any natural risk (e.g., earthquakes, floods, etc.), specifically in regions where there are no precautionary measures, structural protection, and lack to awareness for mitigation measures (World Bank 2017).

## 8.2 Recognizing Natural Hazards

Natural hazards can be well identified and their damages can be observed after they occur, thus post-assessments and surveys are usually applied to deduce the size of impact and to carried out related implements (e.g. medical aids, assure refuge for affected people, removing debris, etc.). However, the foremost implement is to understand the causes and mechanism of the disastrous event. Nevertheless, it can be out of human control to govern them, but it assists taking precautionary measures needed. In other words, when evidences for the occurrence of anticipated disastrous even is identified, that would be very significant. These evidences can play a role for the risk predication including space and time of the natural hazard.

Based on this discussion; however, it is necessary to utilize tools and methods in order to point out for areas under natural risk. This in turn requires recognizing features and indicators that help human to be watch out of risky areas when living these areas.

### *8.2.1 Natural Hazards on Geologic Maps*

Geological maps include several clues that can be used to predicate the occurrence of natural hazards. That is true because these maps are the foremost documents used while applying risk assessment approaches. The immense value of the geological maps enhances the investigator's ability to determine the surficial processes and the related natural hazards. Thus, almost all studies depend on the geological maps to anticipate and assess the existing natural risks and more specifically the vulnerable areas for these risks.

There are tremendous geologic elements can be read on the geological maps. Hence, the successful identification depends on expertise and the number of existing indicative elements on the map. This implies (according to Al Saud 2018a): the geologic distribution of rock lithologies and the lithological characteristics (e.g. rock rigidity, porosity, permeability, etc.); rock deformations, including inclination of bedding planes, and abrupt dip changes between different rock masses, (e.g., folds, flexures, plunging rocks, etc.).

These indicative elements can be summarized as follows:

1. Faults indicators

Geologic maps include the existing geologic structures as an integral part of these maps. The creditability of these maps is often dependent on the used tools and the experts who generated the plotted features.

Faults create surfaces of weakness, and they are always considered as zones of natural risk, this is because many earthquakes take place along fault lines (e.g. San Andreas Fault with 12,000 km). A typical example is the huge earthquake which took place in 1976, along a fault line killed 650,000 people in Tangshan, China.

Another example is from the Red Sea where recurrent and intense seismic activities are frequently reported along the existing faults in the seafloor.

Even though small-scale faults can be dangerous when they are located in detrital rocks or when they are intersected, and thus they form zones with fragmented and loose materials, and terrain instability occurs.

Normally, faults are drawn on geological maps to compose one of the major elements about the geologic structures. However, sometimes faults do not clearly observed on the geological maps because of inaccurate interpretation, or due to the use of inappropriate scales, or sometimes, it is because the existing geomorphologic processes (e.g. alluvial deposition in plains) hide fault lines (Al Saud 2007).

2. Erosional indicators

There are many aspects of erosion which can be directly observed in the field and recognized, but they can be also identified on the geological maps. This ability to identify these processes depends on the volume transported materials and the resulted traces of their pathway. Thus, the pathway along which the erosion takes place is significant to indicate the presence of displaced materials. It can be on small-scale, such as the slide of soil and rock debris along slopes; and it can be with larger scale (i.e. long distances), such as in the case of channel processes where soil and rock materials are transported along watercourses.

Eroded materials evidence the presence of unstable terrain surface as well as they point out to the presence of surrounding tectonic activity that trigger these materials to move. Thus, eroded materials are always shown on the geological maps where they are symbolized alluvial and colluvial deposits, alluvial fans and cones of depression, etc.

3. Volcanic eruption indicators

Volcanics are one of the most dangerous natural hazards and no mitigation measures can be taken to avoid them except the distancing the localities where these volcanics (even the ancient ones) are located. The identification of volcanic eruptions is simply indicated on the geological maps where they are mentioned clearly on the attached legends.

Volcanics are usually related to igneous materials, and more specifically basalt rocks where they often characterized by rounded shapes for volcano craters, but they can be observed as irregular shapes when they relate to lava flow.

4. Coastal indicators

Half of the world's population lives within 60 km proximity to the sea, and three-quarters of all large cities are located on coasts (UNDP 2014). Hence the vulnerability of coastal zones for natural risk is high, and often catastrophic events occur along coasts.

The risk along coastal zones is different than other aspects of natural risks, because to can affected from two opposite sides, it thus can be originated either from the terrestrial or marine side. From the terrestrial, risks can be resulted from erosional processes, collapses, subsidence, flooded depressions, salt domes, etc.

Besides, marine processes are much more damaging, such as erosive action of waves, sea level rise, marine floods, etc. Hence, on geological maps, some of these processes can be identified, such as the abrupt lithological changes at the interface between the marine and terrestrial, sharp sediments accumulation, etc.

### 8.2.2 Natural Hazards on Topographic Maps

The triggering processes into Earth's interior are reflected as features on terrain surface which are often shown on the topographic maps. These features can be revealed in slopes, depressions, and drainages. Therefore, the perturbation in the ordinarily shape and orientation of these features evidences presence of inducing processes, including natural risks. This makes topographic maps significant tool used initially in risk assessment.

Therefore, topographic maps with the orientation and interrelation of contour lines, drainage systems and the plotted features can be used to identify the surficial processes and predict interior movements. Hence, risk-induced features on topographic maps have tremendous indicative signatures including (for example): drainages perturbations, abrupt slopes, irregular altitudes, existed floodplains and depression, rock and soil cones, intermittent coastal morphology, etc.

### 8.2.3 Natural Hazards on Satellite Images

Remote sensing applications have become the most significant tools for studying natural hazards and the monitoring of Earth's dynamics. This wide spectrum of remote sensing largely includes the monitoring and assessment. This can be applied to floods mass movement, forest fire, climatic extremes, desertification, erosion, etc. In addition, forecasting and the early warning systems are mainly based on remotely sensed products along with the geo-information systems.

Therefore, remote sensing is represented by aerial photographs and satellite imageries, and lately drone have become widespread and cheap tool. These tools proved their efficiency in investigating natural hazards including risk assessment and management.

In this regard, remote sensing can be used prior, during and post risk events. This can be diagnosed as follows:

1. Prior-risk event

Remotely sensed techniques are widely used to predict the occurrence of natural hazards. This includes risk pre-assessment, vulnerability and the mapping risk prone areas. In this respect, remotely sensed data are processed to directly identify surficial processes, most of which were mentioned on geologic and topographic maps; hence, zones with obvious risks can be then traced.

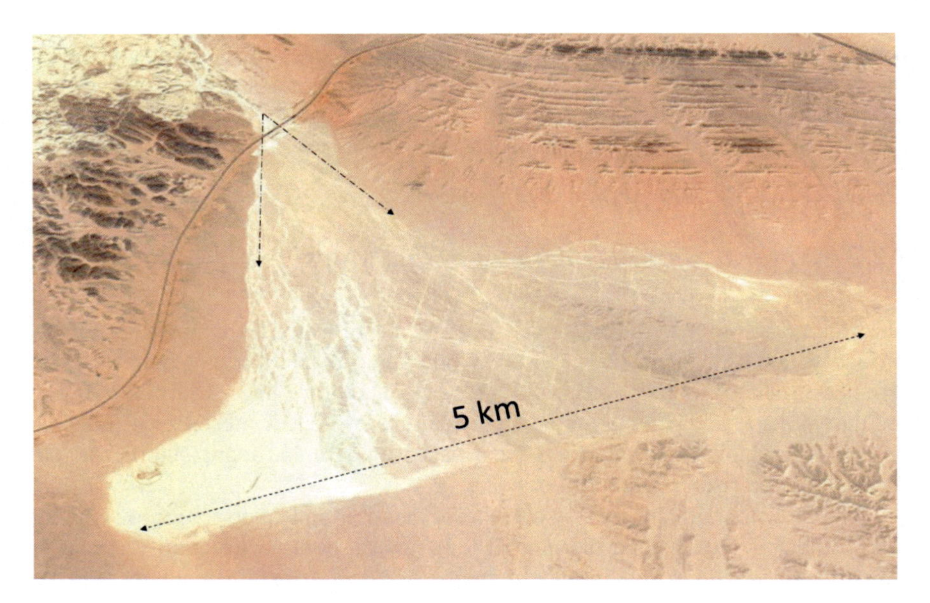


**Fig. 8.1** IKONOS images showing alluvial fan susceptible to the risk of sand encroachment

The direct identification of risk zones, in many cases, requires high resolution images such as Quick-bird (0.61 m), IKONOS (0.82 m, example in Fig. 8.1) where visible band, and sometimes panchromatic are most useful While, some regional processes can be also identified prior the event, and specifically the meteorological events (e.g. dust storm, torrential rain, etc.), where satellite images with short re-visit time, but low or moderate spatial resolution such as MODIS (250 m), Spot-4-Vegetation (1 km), etc.

Other than the direct observation of risk zones on satellite images, these images can be used as a supplementary tool to cartography features which are difficult to be identified on terrain surfaces. A typical example is the detection of faults where thermal images are used to detect the wet horizons that reflecting fault alignments. Thus, the thermal bands on Landsat and Aster images (for example) fit this target.

2. During-event

The use of remote sensing, to monitor the even while it is occurring, is not feasible by satellite images unless the passing time of a satellite is consisted with the event. Otherwise, continuous and prepared surveys or orbiting exist. For such purposes (for example), airborne cameras and Lidar and drones have a significant role.

3. Post-event

The use of remote sensing is also common even after the risk events have taken place, where almost similar satellite images and other remotely sensed techniques can be used as like the case of prior event assessment. This is because direct observations with the visible bands are required. However, the pre-event remote sensing application is usually done to assess the damages including the geographic distribution, degree of impact as well as to deduce factors influenced the occurrence of the natural risk (Al Saud 2010a).

### *8.2.4 Natural Hazards in Records*

Other than mappable and observatory tools to investigate natural hazards, ground measures and registered time series are utmost significant. These include, for example, hydrologic, climatic and seismologic records, as well as the measured data in the field. Therefore, reported numeric measurements are used to identify the trends of measurable processes. Thus, they used in building scenarios and projections required. In addition, records are sometimes used to elaborate the frequency of events, such as in the case of earthquakes recurrence.

Recently, there are several records adopted from remote sensing systems which produce data built on the spectral advantage present in their sensors, such as NDVI, TRMM, CHIRPS, etc.

## 8.3 Seismic Activity

Earthquakes, also described as seismic activities, are tectonic forces originated from sudden energy in Earth's interior and resulted in ground shaking and reflected on terrain surfaces as seismic waves. Thus, earthquakes are mainly caused by the slippage and deformations of rock masses, often along surfaces of weakness (e.g. faults).

Earthquakes account for the majority of deaths in natural hazards which is estimated at 60,000 people a year worldwide—around 90% of them occur in developing countries (OECD 2008). Also, earthquakes, are responsible for approximately 1.87 million deaths in the twentieth century, with an average of 2052 fatalities per earthquake affecting people between 1990 and 2010 (Doocy et al. 2009). While, it was reported that earthquake with 8.9 magnitude occurred in Japan in March 2011, resulted in a tsunami was responsible for more than 28,000 decease. Whereas, an earthquake with 7.0 magnitude happened in Haiti in January 2010 resulted 222,500 fatalities. Thus, the geographic distribution of deaths and injuries resulted from earthquakes differ between regions of different levels of and economic development (Doocy et al. 2009).

There are millions of low-magnitude earthquakes occur annually, but they are not felt by humans (i.e. <2.5 magnitude). Hence, the magnitude of earthquake estimates the amount of energy released, and it often accounts for the damage level. It is described by the moment magnitude scale, which is a logarithmic scale. Thus, a magnitude 6 earthquake is about 10 times less powerful than a 7, and 100 times less than magnitude 8. For example, earthquake with 3.5 magnitude is not harmful, but earthquake with more than 7 magnitude can create widespread damages (Wisner et al. 2008).

Earthquakes are sometime classified according to their type, and the obtained classifications are attributed to the regions where they frequently occur. According to Mitsui (2015), earthquakes can be categorized into four main types. These are as follows:

1. Tectonic earthquakes: These are earthquakes are a result of tectonic plates and occur in Earth's crust by the effect of the geological forces.
2. Volcanic earthquakes: They are earthquakes occurs due to the compression resulted from volcanic activities. Thus, they may followed by volcanic eruptions and lava flow.
3. Collapse earthquakes: These are small-scale earthquake that occurs in regions with underground voids.
4. Explosion earthquakes: They are caused human intervention due to the detonation of chemicals or nuclear devices.

There are a number of earthquake scales adopted to evaluate the magnitude of an earthquake. However, the most common one is the Richter scale.

Seismograms are used to measure the magnitude of earthquakes. They are instruments that measure the motion of ground, including those resulting from the seismic waves, generated by earthquakes, volcanic eruptions and seismic vibration on Earth's crust.

### *8.3.1 Earthquakes: Regional View*

The Kingdome of Saudi Arabia, as a part of the Arabian Peninsula, is vulnerable to seismic activities, and this is well evidenced from the historical records of the Kingdom where several damaging earthquakes occurred. Therefore, the active seismic zones in the Arabian Peninsula have been determined depending of the earthquakes reoccurrence with their magnitudes.

The most intensive tectonic activities are located along the alignment of the Red Sea-Gulf of Aaqba and the Gulf of Aden in the west and south, and extending along Dead Sea, and then the subduction zone associated with the Zagros suture in the north. However, the Red Sea coast is most vulnerable to earthquakes, notably it is bounded by the Arabian Shield, which is a Proterozoic basement rock masses.

The Arabian Peninsula forms a single tectonic plate surrounded by active boundaries where earthquakes often occur (Adams and Barazangi 1984). While, the Dead Sea Transform Fault along the Gulf of Aaqba is the most active region with seismological waves in Saudi Arabia. In addition, there are active transform faults associated pull-apart basins, and hence, representing high damaging earthquakes zone (Al

Damegh et al. 2013). Therefore, the region of the Red Sea Rift System is characterized by tremendous tectonic movements in a broad zone of active deformations between Africa and Arabia.

The most seismic threat for the Kingdom of Saudi Arabia occurs in the Gulf of Aaqba where significant number of earthquakes happen, and many of them were reported over different time periods including the large-magnitude earthquakes such as those occurred in 1983, 1990, 1993, 1995, and 2004 (Hussein and Zaidi 2012).

From April to June 2009, the northwestern Saudi Arabia experienced intense earthquakes which were took place beneath Harrat Lunayyir. The peak seismic activity (19 earthquakes with magnitude exceeding 4) was recorded on 19 May 2009 by the SGS's Telemetric Network of broadband seismometers (Al Saud 2018a). The maximum magnitude recorded was 5.4, and this earthquake caused minor structural damage in the town of Al-Eis (Al Amri and Rodgers 2013).

### 8.3.2 Monitoring Earthquakes

Seismic activities, including earthquakes, are measured using seismographs which are instruments that record seismic waves created by an earthquake, explosion, or other Earth-shaking phenomenon. These instruments are equipped with electromagnetic sensors that convert ground motions into electrical signals, which are processed and recorded by the instruments' analog or digital circuits.

Usually, a number of seismographs fixed to form a full coverage of seismic measurements, and called seismograph networks (SN). The main purpose of SN is the determination of accurate earthquake locations (example in Fig. 8.2). SN is for seismic alarm, or seismic monitoring, and research on the interior of the Earth.

Earthquakes and volcanic activities are monitored by the National Centre for Earthquakes and Volcanoes at the SGS, which also cooperates with the international Earthquake Data Centre for the exchange. Thus, seismograph stations and new networks have been fixed by SGS, and lately they have been upgraded using new broad-band instrumentation and satellite telemetry. Hence, data from remote sites are automatically transmitted via satellite to the SGS processing center.

In addition, seismic data are available the Seismic Studies Center at the King Saud University has established a seismic sub-network at the NW of Saudi Arabia in 1985. It was named the Tabuk sub-network (Al-Arifi and Al-Humidan 2012).

Earthquake epicenter plots are viewed, and magnitudes are calculated. Generally, seismic events within most of western Saudi Arabia are located with an accuracy of just a few kilometers with the present seismograph network. The locations are then plotted automatically, and the data is added to the earthquake database.

About 75% of the stations are concentrated in western Saudi Arabia, where seismicity and risks are the highest. There are about 50 stations are located in Tabuk Province. Eventually, the network will enable the detection and location of earthquakes smaller than the magnitude of 2 anywhere in the KSA (Al Saud 2018a).

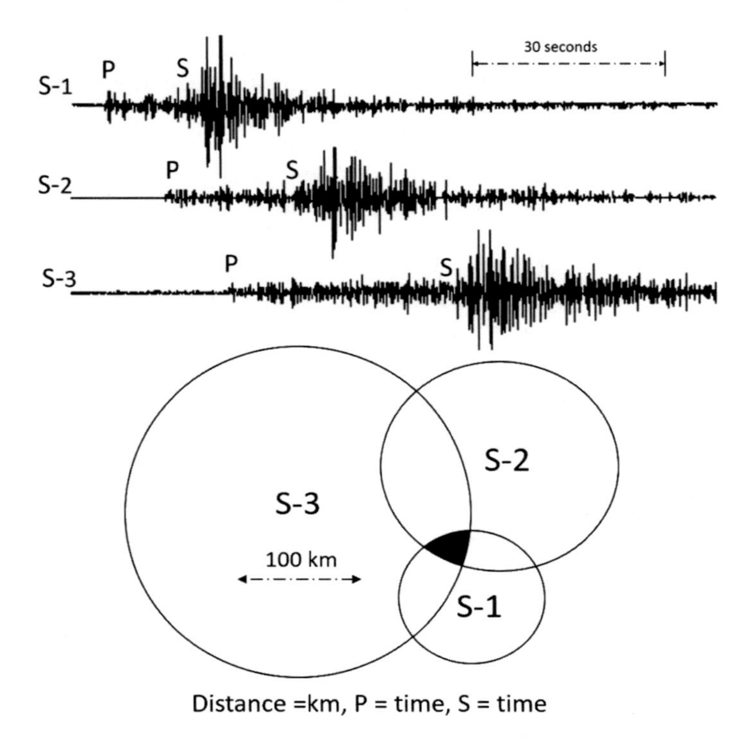


**Fig. 8.2** Location by circle (or arc) method to localize earthquake epicenter by different seismographs

## *8.3.3 Factors on Earthquakes Occurrence*

Seismic activities are internal processes and generated at depth where igneous explosives occur, and thus pushing rock stratum up or down, which is reflected on shelters within the crust of Earth. Therefore, the mechanism of seismic movements is obscure and depends on the energy captured beneath Earth's crust; that is why it cannot be predicted. Hence, the predication of earthquakes occurrence depends mainly on their frequency by region.

In addition, the nature of the terrain where seismic water propagate is significant, and it controls the magnitude of impact. For example, soft materials rocks slightly respond to earthquake vibrations, but it takes large distance propagation, and vice versa with for the consolidated rocks. In addition, the terrain components (i.e., constructions, etc.) are also playing a role in the earthquake intensity and degree of damage.

The following are the controlling on the magnitude and thus damage resulted from an earthquake:

1. Earthquake recurrence: As mentioned previously, earthquakes cannot be precisely predicted, as they occur in different localities at depth due to the unexpected changing energy of the molten rock. Nevertheless, after the occurrence of an earthquake, its source can be precisely identified (example in Fig. 8.2), which is the epicenter, and even the magnitude can be measured. Based on this acquired data on the earthquake, when a region, with define epicenter and even magnitude, witnesses earthquakes frequently, this would be and indicative element used to predict more earthquakes and then assigned this region under seismic risk. Therefore, most studies account for the earthquakes recurrence and then include it as a major factors in mapping earthquake-risk zones (Al Saud 2018a).
2. Rock fractures: The deformation of rocks are significant factors where they play a role of responding after the seismic movements take place. Thus, faults are always representing zone with high vulnerability to seismic activities, notably if these faults are of the active type where fault activity always trigger the terrain surfaces. In addition, faults represent surfaces of weakness in the rock masses, notably if two or more faults are intersected. Therefore, regions with faults differently react with seismic activities than those located in zones with no faults. For this reason, regions along fault alignments are considered as earthquake-prone zones.
3. Lithology: The physical characteristics of a rocks, including consolidation, elasticity and other physical properties such as porosity and permeability, are significant in the propagation of seismic waves through these rocks. Therefore, fragile and soft rock lithologies enhances the transmission of seismic waves for long distances, while this is not similar in consolidated rock lithologies. However, earthquakes that occur in hard and consolidated rocks result higher in magnitude than those that occur in soft rocks. In addition, homogeneity of rock lithology is significant in seismic wave propagation, and heterogeneous lithologies often hinder transmission and sometimes dive the direction of seismic waves (Al Saud 2018a).
4. Terrain components: This factor does not influence the mechanism of seismic activity, but it is related to the degree of earthquake damage. Hence, if an earthquakes with low magnitude occurs in urbanized areas, thus it impact will be much more than it if a higher magnitude earthquake occurs bare regions. Therefore, this factor is related to the vulnerability of a region to and can be used not only for the existing urbanized regions but for the planned urban zones.

### *8.3.4 Tools and Data Analysis*

The identified factors for the existence of seismic activities (i.e., earthquakes) and their impact were manipulated to build an earthquake risk map for NEOM Region. Thus, different tools were used for data analysis and risk assessment. This includes:

1. Seismic records

Data on earthquakes in the region as a whole were collected from local and regional data centers to construct earthquake catalogue (e.g. ENSN Catalog Events 2002; KSU

**Table 8.1** Seismological stations of Tabuk sub-network, which are located in/close to NEOM Region (Al-Arifi and Al-Humidan 2012)

| Station | Code | Latitude N° | Longitude E° | Altitude (m) | Component |
|---|---|---|---|---|---|
| Wadi Mabrak[c] | HQL | 29.27 | 35.05 | 285 | SPZ |
| Haqel[c] | HQL | 29.30 | 34.94 | 005 | SPN |
| Al-Bada'a[a] | BADA | 28.57 | 34.96 | 275 | SPZ |
| Al-Bada'a[b] | BADA | 28.57 | 34.96 | 495 | SPZ |
| Al-Ouyaynah[c] | AYN | 28.87 | 36.00 | 770 | SPZ |
| Al-Sharaf[a] | SRFA | 28.93 | 35.18 | 725 | SPZ,N,E |
| Al-Sharaf[b] | SRFA | 28.95 | 35.11 | 1000 | SPZ |
| Al-Sultaneah[c] | SALT | 29.03 | 34.87 | 350 | SPZ |
| Bir Al-Mashi[c] | BMSH | 28.81 | 34.84 | 050 | SPZ |
| Maqna[c] | MKNA | 28.44 | 34.88 | 650 | SPZ |
| Al-Wajh[c] | WAJH | 26.18 | 36.56 | 75 | SPZ |

[a]Old station, [b]new station, [c]non-moved station

2004; SGS 2010). Moreover, data on the historical seismic records and earthquakes in the region and surroundings are available from Tabuk sub-network where 7 seismic stations are located within/close to NEOM Region. These stations are shown in Table 8.1.

From the earthquakes records, the geographic distribution and earthquakes magnitude were considered and plotted on a map. For the purpose, Arc-Map (in Arc-GIS) was used, and the plotted.

2. Rock fractures map

The map with these geologic features; specifically faults, were extracted from Landsat 7 ETM$^{+}$ and ASTER satellite images, and then the obtained (in Chap. 6, Fig. 6.2) was adopted. It shows all fractures with considerable dimension were plotted. However, three elements representing most significant to be accounted for most seismic activities and much damage can be resulted. These are (according to Al Saud 2018a) as follows:

- Zones with different fracture density; and these can be: (1) more than 30 lineaments/25 km$^2$ for high seismic impact; (2) between 30 and 10 lineaments/25 km$^2$ for moderate seismic impact; and (3) less than 10 lineaments/25 km$^2$ = negligible seismic impact.
- Large-scale faults which exceed 20 km in length; hence, a buffer zone of 1 km from each side of the recognized fault was considered as a seismic-risk zone.
- Faults intersection was also accounted since it represents zones with fractured and unconsolidated rocks.

3. Lithological map

The lithological characteristics, as a main factor in earthquakes, were adopted from the obtained geologic maps for NEOM Region (in Chap. 4, Fig. 4.4). However, the different lithologies on this maps were categorized according to their responding to seismic wave propagation. This has been resulted in 5 major categories where lithologies with earthquake related properties were sorted accordingly.

4. Elements of terrain components

Even though, NEOM Region is almost with very fine urban density, yet some localities where towns and village were considered for the vulnerability to earthquake damage if it happened. These geo-spatial data were adopted from high resolution satellite images (i.e. Spot-7, 1.5 m; WV-2, 0.46 m).

### *8.3.5 Data Manipulation*

For the prepared geo-spatial data on the four identified factors, each factor was elaborated separately through different tools as previously mentioned. However, all these factors were also digitally manipulated for further systematic integration in one board.

Figure 8.3 shows the flow chart for the integration of different factors influencing the occurrence and damage level of earthquakes, which will be produced in a map form. Hence, each influencing factor has different levels of impact. For example, the lithology as an acting factor influencing seismic activity was viewed from the element of lithological elasticity. Therefore, some elements within the identified factors work towards increased occurrence and damage of earthquakes and others are totally opposite, and thus this concept was accounted for data manipulation.

For creditable data manipulation, an empirical classification was done for each influencing factors which were classified into different categories for each (i.e. 5

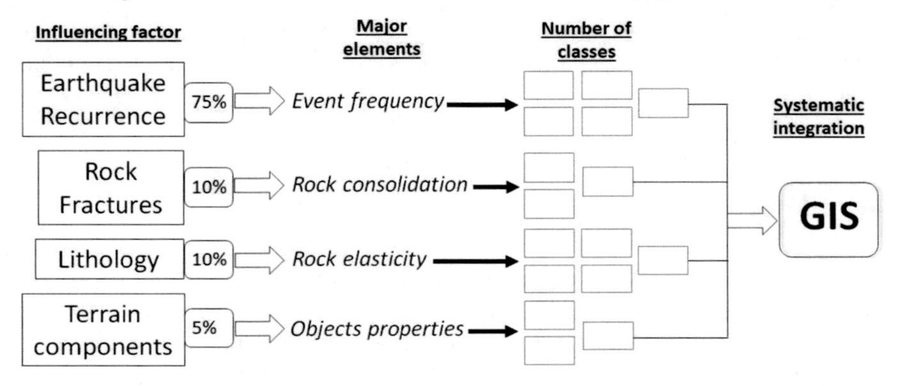


**Fig. 8.3** Flow chart showing the integration of different factors influencing the occurrence and damage level of earthquakes

classes for earthquake recurrence and lithology; and 3 classes for fractures and terrain components as in Fig. 8.3). The geo-spatial data of these factors were converted into maps, and then each map was considered as a geo-spatial GIS layer.

It must be made clear that the influencing factors have different degrees of impact on earthquake occurrence and damage. Thus, some factors of them are characterized as possessing higher influence than others, and this is what should be accounted for during the manipulation of factors. Based on this concept, however, each factor was assigned a defined level of impact, which was described as "weight". In this respect, the following weights were given for the identified factors:

- Earthquake occurrence = 75%
- Rock fractures = 10%
- Lithology = 10%
- Terrain component = 5%.

Following this, reasonable weights for each factor were given. This was elaborated using ESRI's Arc-GIS (i.e., Arc-View) software by overlapping different layers together in the GIS system.

The geo-spatial GIS layers were sorted into 4 major classes, where Class I represents the most vulnerable zones for earthquake occurrence and damage, and Class IV has the least effect. Therefore, earthquakes risk map for NEOM Region was established with four main zones (Fig. 8.4). Thus, the obtained map shows the following:

- Class I: Very high earthquake risk = 3390 $km^2$.
- Class II: High earthquake risk = 2226 $km^2$.
- Class I: Probable earthquake risk = 2253 $km^2$.
- Class I: Moderate earthquake risk = 2857 $km^2$.

There the total areas under earthquake risk in NEOM Region is about 11,026 which is equal to approximately 42%. While the very high and high risk is 5616 which is equivalent to about 21% and this points out that NEOM Region is almost threatened by seismic activity and engineering controls should be accounted.

It is also obvious that the most risky zones are located along the coast of the Gulf of Aaqba, while the southeast part of the region is almost with undefined or negligible earthquake risk.

## 8.4 Floods

For many reasons, water level may rise in rivers, streams and on terrain surface, and thus the excessive amount of water flows chaotically as flooded water and this may cause damages in many regions. Hence, flood is a type of natural hazards which can be mitigated by human if appropriate controls are applied, and this makes is different from other aspects of natural hazards which cannot be controlled (e.g. volcanoes and earthquakes).

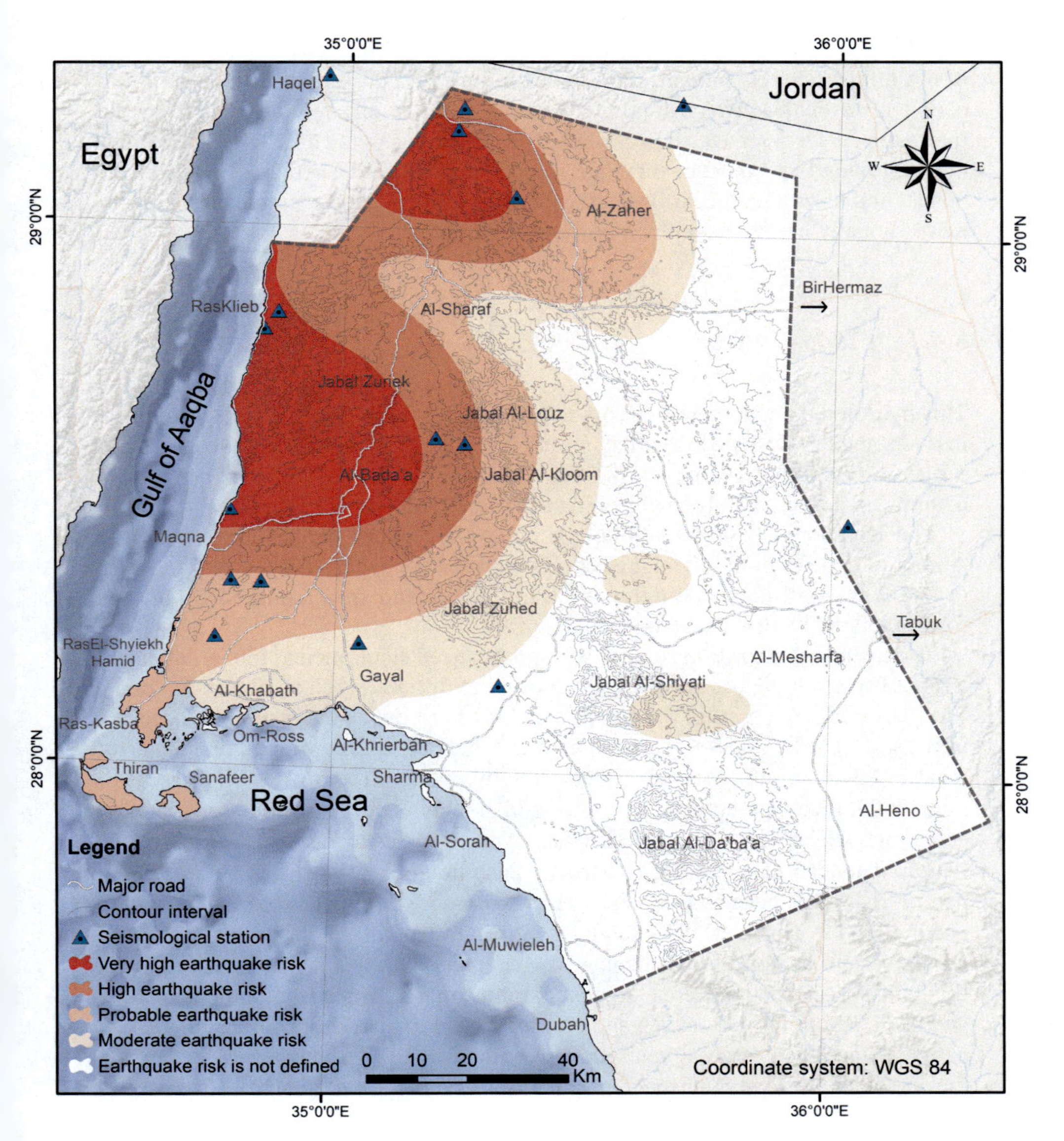


**Fig. 8.4** Earthquake risk map for NEOM Region

When studying natural hazards in many regions worldwide, it always found that the most common type of these disaster events are related to floods. The aspects of water flow on terrain surfaces are controlled by a miscellany of factors which differ from one region to another depending largely on the rainfall intensity and the physical characteristics; in addition, the presence of urban settlements is significant in the level of damages resulted.

Likely to other natural hazards, flood is often resulting losses and destruction in the infrastructures, humans and the environment. Thus, more than one million people are killed each year in poor countries. Hence, a selected time period by the International Disaster Database shows that out of 8 natural disasters, there are 7 belong to flood and most of them are in poor regions (EM-DAT 2020). Therefore, this natural hazard has occasioned considerable concern. However, the problem still exists, notably in regions with no preventive measures.

### 8.4.1 Floods in KSA

This hydrologic phenomenon occurs in wet and dry areas; hence, floods take place in rivers and dry wadis whether the rainfall is regular or torrential and even the flood itself can occur slightly or as flash flood. Therefore, flood has different geographic patterns, including mainly the linear, irregular or geometrical patterns.

The Kingdom of Saudi Arabia, comprising the major part of territory of the Arabian Peninsula, has distinguished geomorphology where large valley systems (>4500 km length) incise different rock lithologies and span a miscellany of land forms including slopes, deserts, mountains, etc.

The historical records of flood is well pronounced since ancient times, and lately a remarkable number of floods have occurred in different regions of Saudi Arabia. They covered large parts of the country, and the most damaging ones were occurred in the last decade, notably in the Riyadh and Jeddah regions which are known for their dense urban settlement that intersect with several wadis.

In 2009, much attention was paid to Jeddah region and the surrounding, the first economical Saudi city which is located along the Red Sea to the west. Therefore, the region of Jeddah has witnessed swarm of flash floods, marking it as an area under natural risk. This became quite evident in November 2009 when torrential rainfall occurred in the urban and commercial localities in Jeddah and its surroundings, and caused serious and unforeseen damage. Later on, in January 2011, another damaging flood took place and covered a larger geographic region. Hence, both floods of 2009 and 2011 resulting in damages and the deaths, disappearance or injury of several people. Since then rarely a year goes by without a flood occurring in Jeddah and the surroundings, and recurrence of this disaster has resulted in an unstable social situation (Al Saud 2015).

The years after Jeddah flood event have witnessed a recurrent number of flood in most cities and regions of Saudi Arabia, including the capital Riyadh. It looks as the entire Arabian Peninsula is entering upon a new climatic epoch. Therefore, rarely a year goes by without swarm of floods strike different regions of the Kingdom. All the reported floods in Saudi Arabia exist as excessive water flow in the valley systems, or water immerse between the alleys and roads. This motivated the concerned sectors to carry out precautionary implementations (e.g. dams, canals, etc.) in many regions of the Kingdom.

The degree of damages results from flood in Saudi Arabia were high. This is attributed to the reason that the region has not been subject to sever flood for long time and this made people unconcerned with this phenomenon, and even the governmental projects were few at that time. Therefore, large human activities were executed in wadis regardless for any expected extreme weather conditions.

For example, the number of people killed in Saudi Arabia in the flood of Jeddah in 2009 is more than the number of people who were killed in Kingdom in different natural disasters combined between 1982 and 2005 (EM-DAT: The OFDA/CRED 2010).This is reflective of the immensity of the adverse impact caused by the flood.

Before, the existed recurrent floods since 2009, there were little studies on floods in Saudi Arabia. Hence, studies existed, they were focused on environmental or geomorphological topics (Al Saud 2004; Qari 2009). Also, while there were studies dedicated to flood assessment, they dealt with the whole of Saudi Arabia (Nouh 1988; Abdulrazzak et al. 1995; Subyani et al. 2009).

The author gave attention to flood topics, and then a number of studies and reports were achieved using space techniques and geo-information systems. The following are selected studies on floods done by the author at different time periods:

- Use of space techniques and GIS to study Jeddah flood (2009).
- Assessment of flood hazard of Jeddah area (2009).
- Mapping flood-prone areas in Jeddah and its surroundings (2010).
- Applying geo-information techniques in studying flood and flash floods in the Jeddah region (2010).
- Use of remote sensing and GIS to analyze drainage system in flood occurrence, Jeddah, Western Saudi Coast (2012).
- Flood control management for Jeddah city and its surroundings (2015).
- Studying the drainage systems for floods in Riyadh Region (2017).

## 8.4.2 Analyzing Factors on Flood

Factors acting in flood occurrence are different between regions. However, the general factors can be as follows:

1. Torrential rainfall where a large amounts of rain falls in a short time period.
2. The unsuitability in the hydrological systems to drain large water volume; and this leads to is the rise in water level in some geographic localities.
3. The underlying rock characteristics and their responses to water flow/infiltration, such as lithology, rock deformations, etc.
4. Distribution of urbanism with respect to watercourses which hinder the uniform run-off in.
5. Human activities, which might result in the restriction of water flow by, for instance, creating depressions where water can accumulate.
6. Lack of precautionary measures (e.g., dams, channels, etc.) to control floods.

For identifying areas under flood risk; there are usually maps for flood-prone area made to point out to the localities under flood risk. These maps, which are utmost significant tools in flood management, also enable the identification the related hydrological controls and processes. Moreover, flood risk maps can be as indicative tool used for implementing flood controls and understanding their dimensions and aspects (Al Saud 2015, 2018b).

The general concept for mapping flood-prone areas implies the determination of the influencing factors which can be a prerequisite step. These factors, once identified, will be systematically integrated together, and each one of them will be given specific weightage in terms of its influence on this natural hazard. This in turn enables the production of a creditable map with different probable risk categories.

1. Rainfall

Rainfall is always considered as the primary and generating factor in floods, and this means that for any region whatever the terrain characteristic are if rainfall does not exist, no flood is anticipated. Thus, the geographic distribution, pattern, and intensity of rainfall are significant in making any area with excessive amounts of water and further on flooding. Hence, if precipitated water exceeds the infiltration rate, flood water accumulates on the terrain surface.

In this regard, rainfall intensity would be more realistic to be considered rather than rainfall rate, because water from regular rainfall, even with high annual rate, may be regularly absorbed by terrain and no chance for flood may be; while excessive amount of water from intensive rain will interrupt the time lag between precipitation and infiltration and thus flood probability exists. The detailed discussion on rainfall in NEOM Region has been mentioned in Chap. 3 (Sect. 3.2.1). Hence, rainfall data were adopted from different sources, notably from TRMM, CHIRPS and other sources (e.g. Meteoblue, GAMEP and World Weather).

2. Drainage system

The involvement of drainage systems in flood assessment is usually analyzed with different approaches where basin geometry and streams morphometry are accounted. Nevertheless, these two main components can be precisely used when flood assessment in relatively separate watersheds is applied. In other words, for a large region like NEOM, the flood assessment (by mapping flood-prone areas) can applied using some hydrologic specifications of basin geometry and streams morphometry which can represent the entire area.

Drainage systems was extracted, from SRTM DEM with a 30 m spatial resolution, with similar approached used in Chaps. 4 and 5. Thus, for NEOM Region, these specifications, which were analyzed for each watershed separately in Chap. 5, are:

- Relief gradient ($R_g$)
- Mean catchment slope ($C_s$)
- Length/Width rations ($Lw_r$)
- Circularity ratio ($C_r$)
- Stream density ($D_d$)

- Meandering ration ($M_r$)
- Texture topography ($T_t$).

Generally, the above hydrologic specifications are function for three major hydrologic regimes. These are the: slope (function of $R_g$ and $C_s$), basin squeezing (function of $Lw_r$ and $C_r$), and stream behavior (function of $D_d$, $M_r$ and $T_t$). Hence, these three hydrologic regimes have significant impact on water flow and thus in controlling flooding processes.

3. Geology

The geology of any area is always a fundamental factor in water flow regime on terrain surface and it involved, in a broad sense, rock lithologies and the existing rock deformations. Therefore, the lithologies were categorized according to their infiltration rate from surface water. That means lithologies with high infiltration rate will absorb much rainwater than those of less infiltration; and thus the first reduces flood probability and the opposite is happened in the second one. Hence, less-infiltration rate lithologies are characterized by high content of argillaceous materials and they often more compacted.

Another component acting on rock infiltration is the secondary porosity and permeability which occurred by when intense deformation are. These are mainly evidenced by the presence of fractures (i.e. faults and fissures) which enhance the rate of water infiltration to substratum.

For the geology of NEOM Region, including the lithology and fractures, they were elaborated (in Chap. 3, Sect. 3.4 and the geologic map, Fig. 4.4) where a detailed description of rock lithologies was illustrated, while the fractures were also plotted on the lineament map (Fig. 6.2).

4. Flood plains and piedmonts

Except studies obtained by Al Saud (2012), rarely a study involved flood plain or piedmont areas in flood assessment; however, these two geomorphological features are utmost significant in a region like NEOM where several wadis from the mountainous regions carry rainwater and transport it along the piedmont fringes where the contact between mountain ridges and coastal zone is.

Hence, it was well pronounced in the field studies that water is always flooded on the existing flood plain which are mainly wide and shallow one. Consequently, the rest water outlets in the coastal zone along the piedmont fringes, and these piedmonts were found to be localities for human activities (e.g. homes, farms, etc.).

5. Human activities

Urban settlements and the related human activities are not only localities of damage from flood, but they are factors making floods. This has been well pronounced lately when several urban sites have been built along the wadis and in localities where water accumulates.

Even though human settlements in NEOM Region are rare enough in the plateau and the mountain ridges, as well as they few in the coastal zone, yet the planned urban sites, as a part of NEOM Project, will be concentrated mainly in the coastal zone where all wadis will flow their water. Therefore, this must be considered for planning approaches, and hence flood controls must be adopted.

### 8.4.3 Flood-Prone Areas

It is the main target, the map of flood-prone areas will be a significant tool in the implementation of SLM for NEOM Region. It is important because the area where flood often occur in NEOM Region is along the coast, the zone where the major projects and activities (e.g. airport, resorts, commercial centres, etc.) will be established.

The production of flood-prone areas map has been applied several times by the author in studied obtained on floods in different regions of Saudi Arabia (Al Saud 2010, 2012, 2015, 2018a). The author; therefore, almost used similar approaches, but some modifications were adopted while modelling the factors creating folds.

The modelling of factors imply manipulation of the geo-spatial data that represented the identified factors (rainfall intensity, drainage systems, geology, flood plains and piedmonts and human activities).

The majority of used tools in this modelling implies the extraction of geo-spatial data and information. Thus, most of these date were prepared in previous chapters as they were mentioned in details. However, the main phases of data extraction includes:

1. Satellite images processing

For identified factors, there are different satellite images used in this regard, including: Landsat (30 m, 60 m thermal), Sentinel-2 (10 m, 20 m SWIN), Spot-7 (1.5, 10 NIR), WV-2 (0.46 m B& W). These images were processed using the software ERDAS-Imagine-2018. Therefore, a number of digital and optical advantages were applied including: band combination, image enhancement, edge detection, directional filtering, color slicing, etc.

The majority of the geo-spatial data required were the identification of wet zones, flooded localities, flood traces (e.g. mud cracks, erosion, etc.). In addition DEMs were extracted from the stereoscopic images where these DEMS are used for slope generation, drainage orientations, etc. (as discussed in Chaps. 4 and 5).

2. Field verification

In accordance with data extraction from satellite images, field checking was carried out consistently to verify the reliability of these data and information. Therefore, filed equipment and devices were used (e.g. Infiltro-meter, GPS, etc.).

3. GIS applications

The consequent step, in flood identification and mapping, was the applications of GIS systems, which served in registering all data and information (from satellite images and field checking) digitally for further geo-spatial data display, manipulation, enhancement, analysis, mapping and digital data storage.

For this purpose, *Arc-GIS,* as the principal Geo-information system tool was used. Thus, the three digital components on Arc-GIS were operated including: (1) *Arc-Map* for geo-spatial data visualizing and performing as well as drawing maps, (2) *Arc-Catalog* for browsing and exploring, creating metadata and managing geo-spatial data, 3) *Arc-Toolbox* for accessing data conversion and analysis.

4. Data analysis and manipulation

Data manipulation was almost the same as in the case of earthquakes, where all identified factors are digitally prepared, and thus each factor with its diverse elements is considered as a geo-spatial GIS layer. These layers would be systematically integrated together to reach the optimal flood map.

For flood-prone areas mapping, the five major factors (rainfall intensity, drainage systems, geology, flood plains and piedmonts and human activities) were digitally overlapped as geo-spatial GIS layers. For this purpose modelling approach is required to presume the influence of each of these factors, notably that each has specific degree of influence. Hence, similarly to the case of earthquakes, each factor was given a weight of impact. However, since flood occurs even in small geographic localities; therefore, weights will be also classified in rates. Whereas, rates represent the degree of influence of the elements within each factor as shown in Table 8.2 and Fig. 8.5. It is clear in Table 8.2 that the number of elements are not the same for each factor, which depends mainly on the presence of effective elements in each factor as well as its areal coverage.

For example, rainfall has relatively wide coverage and it is not restricted for limited area, and thus it has 3 elements, whereas the drainage system treats smaller areas (e.g. sub-catchment) and it has given 7 elements (Table 8.2).

The values applied to the obtained weights and rates were adopted from several sources of studies, most of them were elaborated by the author (Al Saud 2010a, b, c, d, 2012, 2014, 2015, 2018a, b).

5. Modelling and map production:

The applied model has been adopted from Al Saud (Al Saud 2010a, 2015) where the influencing factors have been digitally manipulated in the GIS system, after converting each factor to its percentage of influence, i.e. weight (Table 8.2 and Fig. 8.5). Therefore, each factor has it weight of influence on flood occurrence, while each element has its rate of influence within the factor itself.

For example, geology factor has been weighted at 10% of influence on flood occurrence. However, among this factor there are 4 elements that characterize the geology factor (Table 8.2). Each of these elements also has different rate. Therefore, to calculate the influence of each factor with respect to the total influence on flood occurrence; therefore, a simplified equation must be applied where the weight ($w$) is multiplied by the rate ($r$) for each factor. This should be applied to all factors and their elements. Thus, the sum of all simplified equations for all factors will result a total influence on flood, as follows:

The following equation represents the mathematical components to calculate the sum of effectiveness for factors.

**Table 8.2** Factors on floods and their influence values

| Factors on flood | Elements of influence[a] | Rate (%)[b] | Weight (%)[c] | Total influence (%)[d] |
|---|---|---|---|---|
| Rainfall | Intensive rain | 75 | 25 | 18.75 |
| | Moderately intensive rain | 20 | | 5 |
| | Slight rain | 5 | | 1.25 |
| Drainage system | $R_g$ | 20 | 10 | 2 |
| | $C_s$ | 15 | | 1.5 |
| | $Lw_r$ | 20 | | 2 |
| | $C_r$ | 10 | | 1 |
| | $D_d$ | 20 | | 2 |
| | $M_r$ | 5 | | 0.5 |
| | $T_t$ | 10 | | 1 |
| Geology | Argillaceous | 40 | 10 | 4 |
| | Compacted | 25 | | 2.5 |
| | Partially argillaceous | 10 | | 1 |
| | Permeable | 25 | | 2.5 |
| Flood plains and piedmonts | Shallow flood plain | 35 | 35 | 12.25 |
| | Wide piedmont | 40 | | 14 |
| | Narrow flood plain | 15 | | 5.25 |
| | Regular flood plain | 5 | | 1.75 |
| | Regular piedmont | 5 | | 1.75 |
| Human activities | Activities in watercourses | 35 | 20 | 7 |
| | Activities of flood plains | 25 | | 5 |
| | Chaotic activities | 15 | | 3 |
| | No precautionary measures | 25 | | 5 |
| Total (%) | | | | 100 |

[a]Based on previous studies and field observations
[b]Percentage of weight
[c]Estimations based on author studies
[d]Net degree of influence for each element (rate)

$$\begin{aligned}&\text{Sum of Influence for all factors (Sf)}\\&\quad= \text{Rainfall(Rf)} + \text{Drainage system(Df)} + \text{Geology(Gf)}\\&\quad+ \text{Flood plains and piedmonts(Ff)} + \text{Human activities (Hf)}\end{aligned}$$

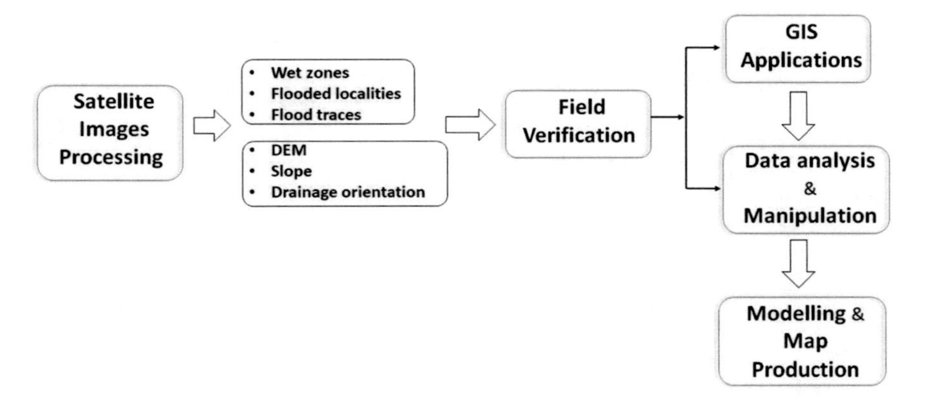


**Fig. 8.5** Model chart showing factors for flood mapping

$$\mathrm{Sf} = \mathrm{Rf}(w\,x\,r)\% + \mathrm{Df}(w\,x\,r)\% + \mathrm{Gf}(w\,x\,r)\% + \mathrm{Ff}(w\,x\,r)\% + \mathrm{Hf}(w\,x\,r)\%$$

Therefore, each element in each factor will have a total influence as out of the Sum of Influence shown in Table 8.2. For example, rain fall has a total weight of 25%, where element 1 (intensive rainfall) has a rate of 75%; therefore the total influence of this element will be:

$$r\% \times w = 75/100 \times 25 = 18.75 \text{ (of the total factor influence).}$$

The results total influence of each element will be systematically converted for the digital applications in GIS system for manipulating all factors together in order to produce the optimal map of flood-prone areas. This can be done by overlapping the 5 geo-spatial GIS layers (for each factor with the obtained element of influence); therefore, the risk map for NEOM Region can be produced showing a number of polygons representing the probability for floods and their damages (Fig. 8.6).

The produced flood map shows that about 1640 km$^2$ of NEOM Region is prone for flood (ranging from very high to high risk), which is equivalent to about 6% of the entire region. Also, it is clear the flood risk is concentrated in the coastal zone, and specifically where piedmonts and wadis outlets exist, as well as flood risk is susceptible to occur along wadis courses. It is also clear that several urban sites including Sharma Runways and the newly existed urban sites are inevitably threatened with flood.

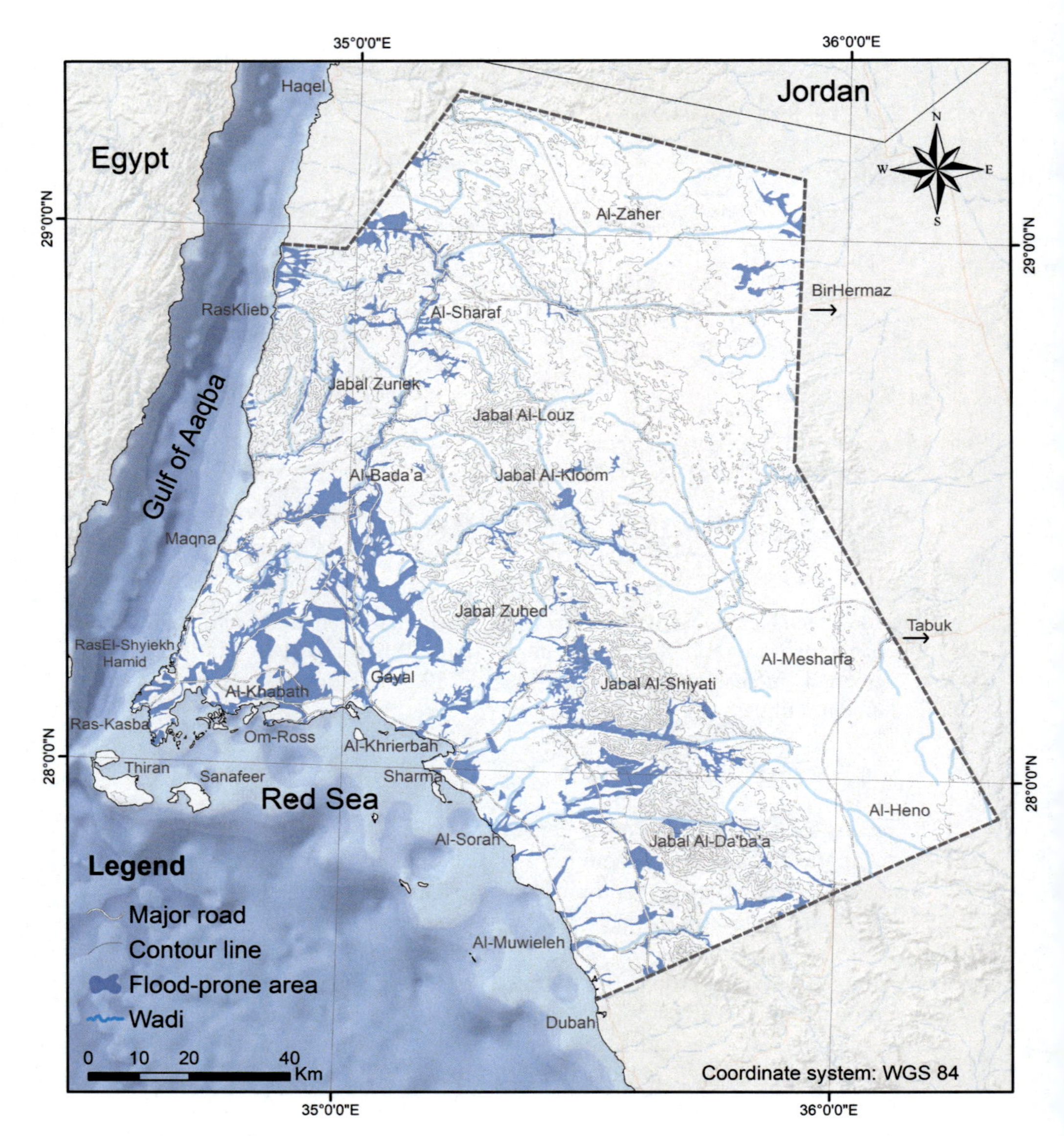


**Fig. 8.6** Flood-prone areas for NEOM Region

## 8.5 Terrain Instability

Terrains with instability often create problems in many regions, notably in poor communities and in regions with urban settlements. Hence, there are many descriptions for terrain instability where the processes of the displacement of materials are usually assigned in different ways including mainly mass movements and erosion. Thus, instable terrain occupies any movement of different terrain materials (rock and soil) where physical agents (e.g. water. slope, wind, etc.) play the main role, and sometimes human activities (e.g. excavation, construction, etc.) can also make impact. In this regards, instable terrain may result unfavorable consequences for the environment and for human as well. The instability of terrain includes different scales that can reach up to several square kilometers.

The mechanism and dimensions of terrain instability plus the geographic distribution of urban sites play significant role in the resulting damages, which can be harmful is some cases. Therefore, terrain stability should be investigated when new human activities (e.g. construction projects, transportation systems, new planned settlements, etc.) are planned. This is in order to deduce terrain vulnerability to any physical process that may lead to movement.

Mapping instable terrain becomes a wide application in many engineering practices and projects, and thus surfaces and processes susceptible to transport materials are always considered. For example, steep slopes represent most hazardous risks surfaces; especially in transportation infrastructure, constructed buildings and to the local economies and environment. Also, earthquake or high groundwater pressures (after heavy downpours) can trigger large rock blocks or even larger assemblages of rocks to crash down on the road surface below (Bunce et al. 1997). All these aspects of terrain instability are usually accounted and the required engineering parties are applied.

### *8.5.1 Aspects of Terrain Instability*

There are several aspects of terrain instability, which differ by region. Therefore, some regions occupy a considerable number of aspects of terrain instability while some another have few; also some region become known by a specific aspect terrain instability. For example some regions are well known by landslides, or subsidence, etc.

1. Erosion

Normally, erosion is a physical process where terrain materials are moved by the effect of water, slope or wind and it may occur over vast area. Therefore, it can be soil or rock erosion and it can be described as hydraulic or aeolian erosion. However, when mentioning erosion, it does not account for the danger results, but it describes the type and mechanism of materials movement. Therefore, erosion can be one of the risk aspects of terrain instability. In other words, the process of materials sliding on inclined surfaces (land-slide) for example is an erosion process, but when it is mentioned as erosion it therefore describes the physical process, and when it is mentioned as landslide, it describe a risk process and so on. Besides, there are several aspects of materials movement on terrain surfaces but they happen without any direct risk; this is represents the general erosion.

In NEOM Region, erosion is well pronounced and it implies the major two types, the hydraulic and aeolian erosion. These have been widely observed in the plateau and the coastal zone, notably where sandstone, loose materials of sands and sediments exist. Hence, lately occurred erosion can be over vast areas where eroded materials exceeds 1 m in many places (example is shown in Fig. 8.7).

2. Rock fall

If any terrain includes rock falls, it is therefore considered as instable terrain and it will be designated as an area under natural risk, but this happens at local scale (i.e. limited geographic patches). Hence, rock fall (or mass wasting) always attributed to the movement of surface materials on air, which means no considerable movement is happened along the terrain surface, and this is the difference with landslide where the movement should be along terrain surface.

Alternatively, the movement can also be in intermittent contact with the terrain, such as those seen in rock crumpling, toppling, and rolling. Hence, the mechanism of material movement in this type of rock fall is usually found to be a very fast occurrence. This movement is also governed by factors similar to those in landslides, but the surface materials in this case are made of almost hard and consolidated rock

**Fig. 8.7** Eroded sandstone along the road between Bir Hermas and Sharma (the estimated erosion thickness is about 1.2 m)

of different dimensions—ranging from small gravel and debris to rock boulders (Al Saud 2018b).

In NEOM Region, rock fall has been well observed in the area of mountain ridges where hard, massive and fractured rocks are dominant, as well as the acute sloping surfaces. Therefore, mass wasting, toppling and crumbling are widely found.

3. Landslides

Landslides are common phenomenon of instable terrain where soft and hard materials of rock and soil move along sliding surfaces. These aspect of natural hazard are usually occur locally and within a limited geographic area. Thus, the damage results from landslides is often local and it depends mainly on the dimension including the volume of moved materials. While, the mechanism of movement is controlled by many factors including mainly the surface angle and roughness, type of transported materials, as well as acting factors such as seismicity, water and human influences that trigger these materials.

The movement speed of landslides can be from creeping to flash movement where the latter results the damage which is almost local. Hence, landslides almost redistribute rock debris, soil and sediments through fast collapses and sliding. Therefore, several types of landslides are known, specifically are: rotational, translational, debris flow, earth flow, slope creeping, and lateral spread type.

Landslides, with relatively large dimensions, can be noticed in NEOM Region. This is mainly found in the coastal zone, and specifically along the contact between mountain ridges and the coastal plain where there soft lithologies overlying massive and hard rocks.

4. Subsidence

This type of terrain instability, which is usually associated with cracks and fissures, occurs through the pull-down (subsidence), and it may be accompanied with pull-up (uplift) of the terrain surface, and then forming wavy terrain. However, it also occur as separate concave down in terrain level due to many reasons including mainly: ground cracks, faults intersection, caves and conduits and other aspects of karstification, change in soil nature, excessive abstraction of groundwater, loess soil and seismic activities.

Even though, subsidence is well pronounced even in the desert areas, like in Saudi Arabia, yet the resulted earth fissures and subsidence could cause many problems in different urban and agricultural areas and induce damages to the infrastructure (Holzer 1984).

In NEOM Region, subsidence was noticed in the coastal plains and more certainly along the shoreline where loose materials ad invaded by seawater. The observed subsidence are expected to be more than what has been observed and they need more investigations to be identified as they form a treating agent for the coastal urbanism.

### *8.5.2 Factors on Terrain Instability*

There are several factor acting on the occurrence of terrain instability where they can be physical, anthropogenic or combination of both factors. These factors are controlled by the components exit in the region where they develop, and thus resulting natural risk at different levels. Therefore, the majority of these factors can be summarized with the following factors:

1. Climate: Climate as physical factor plays a significant role in terrain instability where the rain and temperature trigger the rocks to move. Thus, rainfall, as water drops of running water, is a generating factor where it either directly affects terrain materials (i.e. weathering, solubility, etc.), or it is acting in pushing these materials. While temperature changes the interior structure of surface material from one side, sand it destructs (e.g. shrinking, expansion, etc.) it in the other side.
2. Morphology: The morphology, as an integrated factor with geology, represents the zone along which terrain instability occurs, thus the characteristics of terrain morphology is significant in this regards. However, these characteristics have a number of elements where the fundamental ones are the slope and surface roughness. Therefore, surfaces with steep slopes and smooth surface are much vulnerable for terrain instability and vice versa.
3. Geology: This is a function of rigidity and consolidation terrain materials where it happens due to many physical elements, such as seismic activities, weathering, etc. Hence, the presence of poorly consolidated, friable, porous, permeable and weathered materials are much more susceptible to be instable, and these properties are controlled by the lithological characteristics of rock (including top soil) as well as the existing structures (e.g. fractures, joints, etc.) on these lithologies.
4. Human interference: Usually the negative human intervention trigger the probable physical processes, such as in the case of floods, and here in the stability of terrain. This is common factor that destruct the confined and stable materials. It includes several aspects of human interference, such as excavation, mining, groundwater abstraction, etc. Nevertheless, human interference cannot be accounted while analysing all factor together because it is usually occurred locally with no define timing or place (Al Saud 2018).

These factors can sometimes act separately (i.e. each factor develop the instability of terrain and mass movement), but in many case they accumulate the energy of impact unless the instability takes place, such as a material on slopes which are susceptible to slide, but if a human makes any work in the context of this slope, hence, this material will slide.

Moreover, all these factor are usually interfered and have the same origin or they may work along similar processes. For example, seismic activities controls the degree of fractures in rocks where the latter will be instable materials along smooth surfaces which can be easily moved when a rainfall event takes place.

### 8.5.3 Analysing Factors on Terrain Instability

The analysis of the acting factors on terrain instability in this documents implies the principal elements of these factors and the selection of these elements based on the expertise on the subject matter as well as the field observations. Therefore, these elements were diagnosed and analysed using the suitable and available tools where satellite images and geo-information systems were the main analysing tools along with field surveys.

1. Rainfall: Similarly to data retrieve in flood assessment, rainfall records were illustrated using TRMM and CHIRPS plus other rainfall data sources and ground measurements. Consequently, rainfall was classified into 4 categories. These categories are: intensive rain, moderately intensive rain, slight rain and fine rainfall.
2. Terrain slope: This also followed the same approaches of slope extraction which was used in Chap. 4 (Sect. 4.7.1) where SRTM DEM was processed using Arc-GIS software (Toolbox system). Therefore, similar classification was also adopted as follows: $<5°$, 5°–10°, 10°–20°, 20°–30° and $>30°$, and the same was also used (Fig. 4.7).
3. Rock fractures: Identification of rock fractures in NEOM Region was dependant on the lineament map (Fig. 6.2) and the field observations. This map which was established from the processing of Landsat 7 $ETM^+$ and Aster images, were classified to fracture density exactly as performed in Chap. 6. Thus, the density was classified into 5 classes as: $>40$ lineaments/25 $km^2$ for the very high density up to $<10$ lineaments/25 $km^2$ for the very low density as in Table 6.2.
4. Lithology: Lithologies with similar properties towards terrain instability were sorted from the 80 identified rock formations in NEOM Region. Therefore, the elaborated geologic map (Fig. 4.4) for NEOM Region was used. These rock formations, which were described by Rowailhy (1985), Clark (1987) and by Davies and Grainger (1985), were categorized into four classes as follows: argillaceous, compacted, partially argillaceous, consolidated and massive rocks.
5. Landforms: It is a function of the topographic characteristics where different landforms reflect terrain vulnerability to instability of the existing materials. In this view, the obtained landform maps (Fig. 4.5) was used. Therefore, landform classes illustrated in Table 4.3 were categorized into 5 classes as follows: alluvial deposits and sand, coastal deposits, hills and outcrops, mountains and lava and volcanic hills.

### 8.5.4 Modelling Factors on Terrain Instability

Factors manipulation was applied after converting each factor with its elements into geo-spatial GIS layer. That is similarly to the modelling method applied for flood assessment where each factor will represent a digital layer with several elements prepared for further integration of all factors in order to produce the final terrain instability map. In addition, each factor was given a weight and the elements were given rates. Thus, a detailed diagnose of 5 factors with 23 elements.

Therefore, weights are rates were put depending of previous applied work by the author (e.g. Al Saud 2018) as well as the field observations and surveys applied in the region. The same modelling used in flood assessment, was used here where weights and rates are multiplied and then the resulted values were accounted for the total influence on terrain instability.

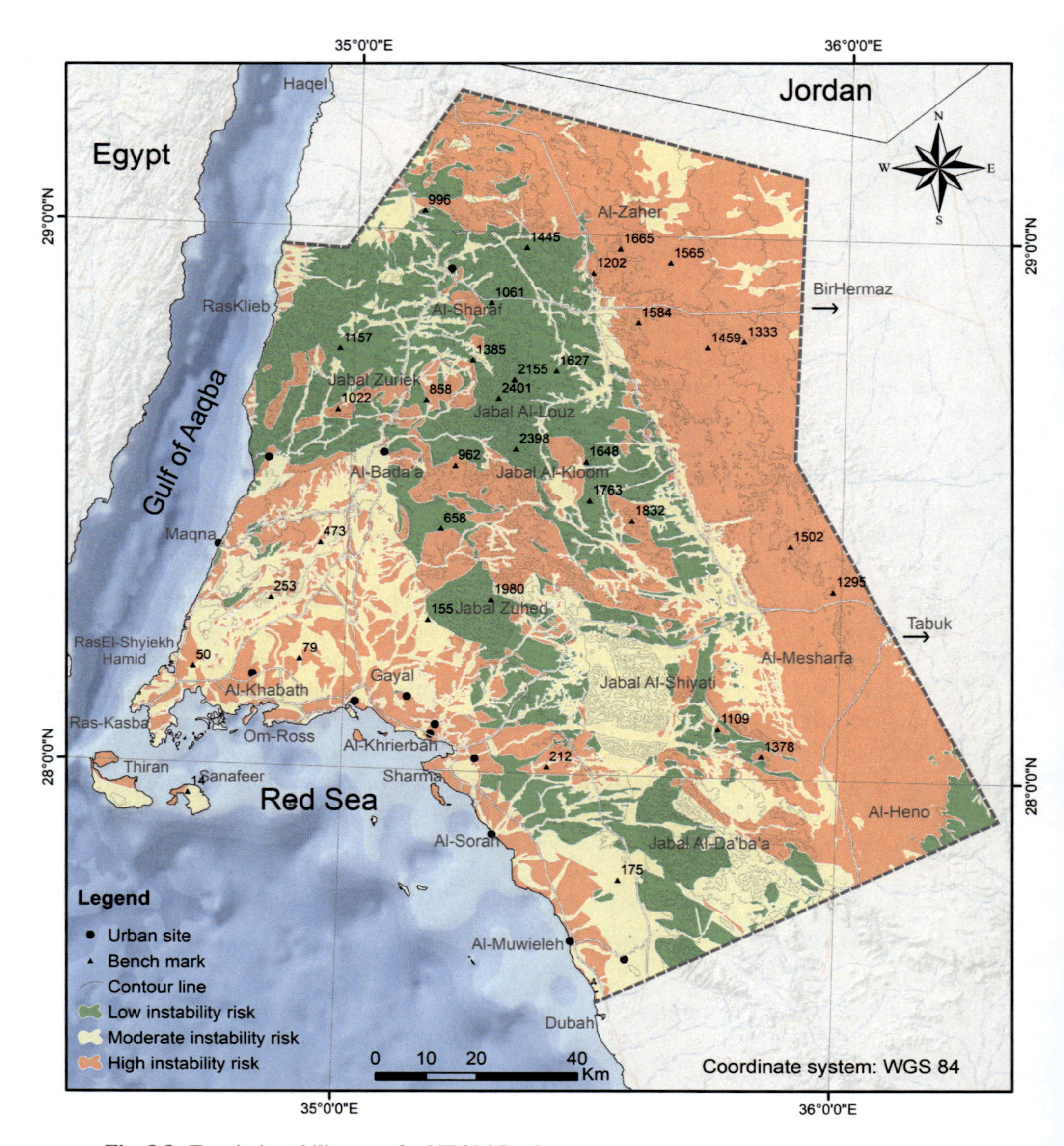


Fig. 8.8 Terrain instability map for NEOM Region

Further on, total influence of each element was systematically converted for the digital applications in GIS system where all factors will be integrated together in order to produce the optimal map with three major categories for terrain instability (Fig. 8.8).

Therefore, the produced map shows there are 14,562, 5753 and 6185 km$^2$ for the areas with high terrain instability risk, moderate and low; respectively. It is obvious that the instable areas are concentrated in the plateau where movable sand and sandstone are dominant, and then in the coastal zone where loose materials are widespread along the piedmonts, and the shoreline area.

## References


Abdulrazzak, M. J., Sorman, A., Onder, K., & Al-Sari, A. (1995). Flood estimation and impact: Southwestern region of Saudi Arabia. King Abdulaziz City for Science and Technology; Project No. ARP-10-51, Riyadh. Saudi Arabia.

Adams, R. D., & Barazangi, M. (1984). Seism tectonics and seismology in the Arab region: A brief summary and future plans. *Bulletin of the Seismological Society of America, 74*(3), 1011–1030.

ADPC (Asian Disaster Preparedness Center). (2000). *Capacity building in Asia using information technology applications. Concepts of hazards, disasters and hazard assessment. Course material. Module 2.* Available at: http://www.adpc.net/casita/course-materials/Mod-2-Hazards.pdf.

Al Amri, A. M., & Rodgers, A. J. (2013). Improvement of seismicity parameters in the Arabian Shield and platform using earthquake location and magnitude calibration. In K. Al Hosani, F. Roure, R. Ellison, & S. Lokier (Eds.), *Lithosphere dynamics and sedimentary basins: The Arabian plate and analogues* (pp. 281–293). Berlin: Springer.

Al Damegh, K. S., Abou Elenean, K. M., Hussein, H. M., & Rodgers, A. J. (2013). Source mechanisms of the June 2004 Tabuk earthquake sequence, Eastern Red Sea margin, Kingdom of Saudi Arabia. *Journal of Seismology, 13*(4), 561–576.

Al Saud, M. (2004). Study of environmental problem in Jeddah as a big City in the Kingdom of Saudi Arabia and the methods of facing it, *Saudi Geographical Society.*

Al Saud, M. (2007). Using satellite images to study drainage pattern anomalies in Saudi Arabia. *Journal of Environmental Hydrology, 15*, 1–14.

Al Saud, M. (2010a). Assessment of Flood Hazard of Jeddah Area 2009, Saudi Arabia. *Journal of Water Research and Protection (JWARP), 2,* 839–847.

Al Saud, M. (2010b). Application geo-informatics techniques in the study of floods and torrents in Jeddah in 2009. *Arab Journal of Geographic Information Systems, III*(1) (Arabic edition).

Al Saud, M. (2010c). Map flood risk and floods in Jeddah. *Geographical Research Journal, Saudi Arabia, 91* (Arabic edition).

Al Saud, M. (2010d). Use of space techniques and GIS for mapping transported sediments: The case of Jeddah Flood 2009, Saudi Arabia, World water Week in Stockholm, September 5–11.

Al Saud, M. (2012). Use of remote sensing and GIS to analyse drainage system in flood occurrence, Jeddah—Western Saudi Coast (Chapter) In *Book entitled: Drainage Systems. InTech Open Science* (pp. 139–164). Croatia. ISBN13: 978-953-51-0243-4.

Al Saud, M. (2014). *Mapping flood-prone areas in Riyadh Region (In Arabic). The high commission for the development of Riyadh.* Series of 20 Sheets and Technical Reports (including GIS, maps). 422 pp.

Al Saud, M. (2015). *Flood control management for the city and surroundings of Jeddah, Saudi Arabia* (177 p). Springer. Heidelberg, New York. ISBN13: 978-94-017-9660-6.

Al Saud, M. (2018a). *Using space techniques and GIS to identify vulnerable areas to natural hazards along the Jeddah-Rabigh Region, Saudi Arabia* (306 p). New York: Nova Science Publisher Inc.. ISBN13: 978-15-361-33134.

Al Saud, M. (2018b). *Geomorphological characteristics and flood-prone areas of the Wadi Al-Saly Basin, Riyadh Region* (138 p). Boston: Publishing Solutions Group. ISBN13: 978-1-9561-1.
Al-Arifi, N., & Al-Humidan, S. (2012). Local and regional earthquake magnitude calibration of Tabuk analog sub-network, Northwest of Saudi Arabia. *Journal of King Saud University-Science, 24*(3), 257–263.
Bunce, C. M., Cruden, D. M., & Morgenstern, N. R. (1997). Assessment of the hazard from rockfall on a highway. *Canadian Geotechnical Journal, 34,* 344–356.
Burton, I., Kates, R. W., & White, G. F. (1978). *The environment as hazard New York*. Oxford: Oxford University Press.
Clark, M. (1987). Geologic map of Al-Bada'a quadrangle, A-28; (1:250.000). Ministry of Petroleum and Mineral Resources.
Davies, F., & Grainger, D. (1985). Geologic map of Al-Muieleh quadrangle, A-27; (1:250.000). Ministry of Petroleum and Mineral Resources.
Doocy, S., Daniels, A., & Aspilcueta, D. (2009). Mortality and injury following the 2007 Ica earthquake in Peru. *American Journal of Disaster Medicine, 4*(1), 15–22.
EM-DAT (International Disaster Database). (2020). *Centre for research on the epidemiology-CRED*. Available at: www.emdat.be/.
EM-DAT: The OFDA/CRED. (2010). *International disaster database, Université catholique de Louvain, Brussels, Bel*. Data version: v11.08. www.emdat.be/.
ENSN Catalog Events. (2002). Seismic data for the western Arabian Peninsula 1956–2002.
Hewitt, K., & Burton, I. (1971). *The hazardousness of a place: A regional ecology of damage events*. University of Toronto.
Holzer, T. L. (1984). Ground failure induced by ground-water withdrawal from unconsolidated sediment. In T. L. Holzer (Ed.), *Man-induced land subsidence* (Rev Eng Geol, VI, and the Geological Society America) (pp. 67–105).
Hussein, M. T., & Zaidi, F. K. (2012). Assessing hydrological elements as key issue for urban development in arid regions. In S. Polyzos (Ed.), *Urban development* (pp. 129–158). Rijeka, Croatia: InTech.
KSU. (2004). Seismic data for the western Arabian Peninsula 2002–2004, King Saud University, KSA.
Mitsui, Y. (2015). Types of earthquake sources. *Earth, Planets and Space, 67*(1).
Nouh, M. A. (1988). Estimation of floods in Saudi Arabia derived from regional equations. *Journal of Engineering Science, 14*(1), 1–26.
OECD (Organization for Economic Cooperation and Development). (2008). *Costs of inaction of environmental policy challenges Report ENV/EPOC (2007)17/REV2.*
Qari, M. (2009). Geomorphology of Jeddah Governate, with emphasis on drainage systems. *Journal of King AbdulAziz University (JKAU): Earth Science, 20*(1), 93–116.
Rowailhy, M. (1985). *Geologic map of Haqel quadrangle, A-29; (1:250.000).* Ministry of Petroleum and Mineral Resources.
SGS. (2010). Seismic data for the western Arabian Peninsula 2005–2010, Saudi Geologic al Survey, KSA.
Subyani, A., Qari, M., Matsah, M., Al-Modayan, A., & Al-Ahmadi, F. (2009). *Utilizing remote sensing and GIS technologies to produce hydrological and environmental hazards in some Wadis, Western Saudi Arabia (Jeddah-Yanbu).* King Abdulaziz City for Science and Technology: Department of Hydrology. General Directorate of Research Grants Program. Kingdom of Saudi Arabia.
UNDP. (2014). *Human development report 2014. Sustaining human progress: Reducing vulnerabilities and building resilience.* UNDP, New York.
Wisner, B., Blaikie, P., Cannon, T., & Davis, I. (2008). *At risk: Natural hazards, people's vulnerability and disasters* (2nd ed., p. 275). New York: Routledge.
World Bank. (2017). *Natural disasters in the Middle East and North Africa: A regional overview. Documents and reports.* Available at: http://www.worldbank.org/en/region/mena/publication/natural-disasters-in-the-middle-east-and-north-africa.

# Chapter 9
# Compatible Land Management

**Abstract** In the previous chapters, a detailed discussion has been carried out on several themes which are necessary to be highlighted before allocate the major elements of SLM for NEOM Region. Without identifying these themes, including their geographic dimensions, the application of SLM will not be under the right way, and therefore any proposed activity related to the development for this region will be inaccurate and may result unsuccessful management. This has been well pronounced in many regions worldwide including the Kingdom of Saudi Arabia. Therefore, the author aimed at introducing the necessary database and highlight on the fundamental issues related SLM, based on advanced scientific methods and tools. Hence, the management approaches in the region of NEOM should follow hierarchy and consistency with the nature of the region. Therefore, the principal physical characteristics and maps plus other terrain properties were primarily determined as background data and information needed. This has been followed by identifying the available resources and the threatening physical processes, i.e. natural hazards. Thus, all components for the optimal SLM were therefore prepared in this document in order to end up with this chapter where the major land management implementations and activates would be the compatibility for successful management. This chapter will present five major compatible land management components including: urbanism, transportation, entertainment, heritage and archeology and the green energy.

**Keywords** SLM · Railway · Resorts · Heritages · Green energy · Smart city

## 9.1 SLM Indicators for NEOM Region

It is no longer acceptable that the economic growth and countries development are being done on the account of degrading the natural resource and the deterioration of environmental quality. It is no less urgent is the need for better and more environmentally land management and nature protection.

No doubt, land provides significant environmental and even life-need benefits, such as its role in the biochemical cycles (water, carbon, oxygen cycles etc.), water de-pollution, as a source and sink functions for greenhouse gases, etc. The challenge


M. M. Al Saud, *Sustainable Land Management for NEOM Region*,
https://doi.org/10.1007/978-3-030-57631-8_9

remain on how to utilize this nature and how to protect it in order to sustainably continuing its productivity for human being. That is not a utopian dream, but evidences of land quality and SLM are needed to lead us along the way.

Indicators are descriptions that represent a status and convey information on the changing trends of that status. Thus, indicators may be used in monitoring and evaluation programs to estimate the rate of change and the impacts. Indicators are already in regular use for economic and social data. While, there are few indicators available to assess, monitor, and evaluate changes in the quality of land resources or the impact of human interferences on the environment (Dumanski et al. 1998).

There are several indicators put for the sustainability where many of them were assigned for different themes such as for land use, agronomical contamination, etc. These indicators were used to regulate the trends of implementing these themes. Thus, indicators are monitoring tools to identify the progress of the managed themes. It is; therefore, a necessary tool for SLM and it can be done prior and after implementing SLM for any region like NEOM, especially indicators will guide the actions taken for the development and the sustainability of these actions. Therefore, the selected SLM indicators to be adopted for NEOM Region can be represented in Fig. 9.1.

It is; therefore, obvious that three major aspects of SLM indicates can be elaborated for Neom Region, including the economic, environmental and social indicators, where the economic one should be viewed from the integration of national and international financial and commercial resources, while the ones for environmental and social are applied mainly for NEOM Region as a selected global hub.

The next step is the identification (or selection) for the needed indicators which can be grouped (i.e. a matter of database inventory) for further data organization and integration between different selected indicators (Fig. 9.1). Consequently, the selected

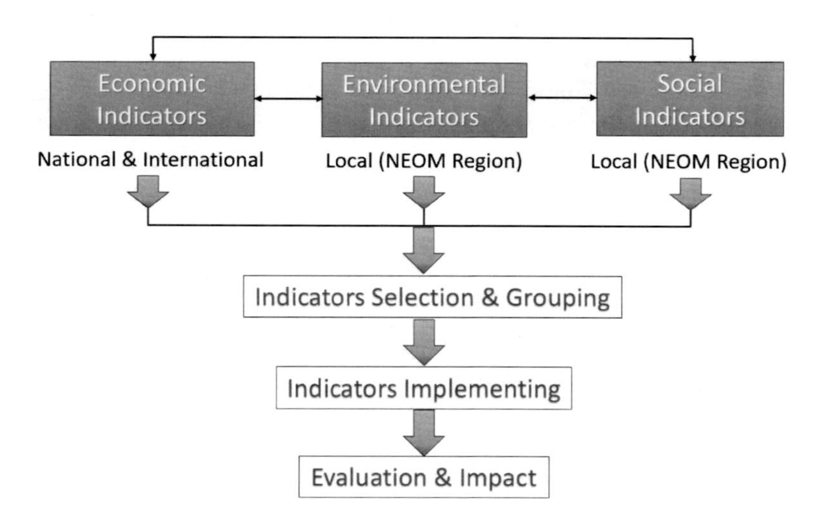


**Fig. 9.1** Selected SLM indicators for NEOM Region

indicators should be implemented for NEOM Region by establishing monitoring and follow-up tools, which will be discussed (below) for each indicator separately.

Eventually, evaluation and impact assessment of the implemented indicator must be carried out in order to find out whether these indicators were applied correctly or they need to be regulated according to any existing condition. In addition, the impact of implementing these indicators will be evaluated for further modification to reach the best orientation of SLM in NEOM Region.

1. Economic indicators

Economic indicators are usually given more attention. Thus, they are represented by numeric assessment and statistics about the economic activities for analysis of economic performance and predications. They include various indices, earnings reports and inventories. Thus, economic indicators are classified into two main categories. These are: (a) leading indicators which represent the indices that change before the entire economy is changed and (b) lagging indicators are generally those indices which change after the entire economy has changed. Hence, the main economic indicators (but not limited) are:

- Income loss and per capita income,
- Economic burden,
- Employment opportunity,
- Interest rates,
- Gross Domestic Product (GDP),
- Government regulation and fiscal policy,
- Existing home and properties sales.

Since NEOM Region is sought to be as international commercial center with many trading and project clusters (e.g. industrial zones, banks, etc.); therefore, the economic aspect is significant not only for foreign investors, but also for the locals. Hence, the economic part for NEOM project must be under control and all commercial and economic trends should be continuously elaborated. This needs integrated deal with different stakeholders (also foreign and local).

Therefore, it is necessary to execute an economic monitoring system dedicated for NEOM Region. This can include (partially adapted from AAII 2020):

- Applying different economic indices (e.g. Consumer Price Index, CPI; Producer Price Index, PPI),
- Preparing periodical economic inventories,
- Monitoring stock and labor market data,
- Establishment of domestic and international financial centers (e.g. banks, insurance companies, etc.),
- Current Employment Statistics (CES),
- Consumer Confidence Survey,
- Money Supply (M2) which calculates the money supply including cash and checking deposits, money market, mutual funds, etc.

2. Environmental indicators

These are simple measures that evidence and inform about the processes which are taking place in the environment. As a result of environmental complexity, these indicators provide a more practical and economical way to monitor and trace the state of the environment rather than trying to investigate every possible variable in the environment. The following are the most adopted environmental indicators (almost examples) which can be applied to SLM. They were mentioned in many sources such as Hammoun et al. (1995) and Sam et al. (2016).

- Species extinction and endangered species,
- Different ecological footprint,
- Land use/cover change,
- Changes in water bodies (dimensions and quality),
- Land degradation and desertification,
- Quality deterioration,
- Acidification and eutrophication of the Environment,
- Dispersion of toxic Substances,
- Groundwater depletion,
- Increased undisposed wastes (solid and liquid),
- Loss in ecological natural resources,
- Changing climate (e.g. rainfall, temperature, etc.).

For NEOM Region, the environmental indicators are utmost significant and should be under continues follow-up and monitoring. This is because the environmental components of this region represent one of the most important components that assigned it to be a global hub with a miscellany of natural features. For this purpose, all investigation, assessment and monitoring tools should be available while NEOM project is started. This will meet with the requirements for elaborating the mentioned indicators. Therefore, the following tools are required:

- Fixing field instruments (e.g. gauges, installed devices, etc.),
- Laboratories and research centers,
- In situ testing tools (i.e. portable measuring devices),
- Early Warning Systems (EWSs),
- Forecasting (e.g. mainly for weather and climatic extremes, etc.),
- Observatory stations (e.g. satellite imageries, drones, etc.),
- Sharing global networks (e.g. seismic networks, etc.),

Elaborating environmental legislations.

3. Social indicators

Social indicators measure the degree of development apart from any financial or economic means. However, they are indirectly related with the economic and even environmental aspects of development. Therefore, the social aspect in the development is usually given concern by the inhabitants who are (or their properties) involved in the development plans. Hence, the most adopted social indicators are:

- Human health,
- Livelihood,
- Cultural heritage,
- Local participation and acceptance,
- Trends on future generation,
- Impact on property,
- Local food supply chain.

NEOM Region, with fine population density, is expected to gather a considerable number of people from outside the region whether these people are from Saudi Arabia or from abroad. However, the social indicators would be evaluated mainly according to the respond from the Saudi people first and then the integration of people from other citizenship.

In order to elaborate social indicator for NEOM Region, the following implementations can be applied, even though these more implementations can be developed according to the existed social components and demographic changes:

- Evaluation of Illiteracy level,
- Evaluation of cultural and educational level,
- Surveying the level of social problems,
- Crime rating and acts disturbing security,
- Population growth rates (births, deaths, etc.),
- Heritage and archeological conservation,
- Evaluating prosperity level and satisfaction in the society,
- Engagement of locals in the executed projects,
- Awareness and educational campaigns,
- Establishment of cultural center and national libraries,
- Securing jobs for the local population.

## 9.2 SLM Compatibility of NEOM Region

Land management compatibility must be promoted through proactive planning and land classification. It is to investigate whether the natural components are compatible to apply specific urban activities (e.g. construction of urban settlements, building rods, etc.).

Up to date, no compatibility framework has not been elaborated in-depth for Neom Region in order to insure appropriate compatibility for land management. Thus, a number of implements should be considered for this purpose. These are:

1. Projects should comply with national policies and strategic plans, otherwise, constraint will exist.
2. Gross land use compatibility with land use typology area should be assessed (e.g. traffics, building scale and site design, operational impact, etc.). This need detailed surveys to determine whether or not the proposed project may have a conflict with existing users.

3. Defining the naturally risk and hazardous zones in order to avoid the risky places or apply precautionary measures.
4. Identifying zones with natural resources (geological and ecological resources). This is essential to implement plans for investment from one side and to conserve these resources from the other side.
5. Zoning heritage and archeological sites for protecting these sites and to involve them in the urban planning schemes.
6. Infrastructures and energy sources availability (e.g. sewage systems, power plants, pipelines, etc.). Thus, they can be connected with the proposed projects if they exist or plan for new infrastructures in case they are absent.
7. Geography (domestic and internationally) of the proposed projects area with respect to other regions, including international transport accessibility.
8. Climatic conditions, with as special emphasis on climatic-induced risks, such as areas with frequent snowing, or dust storms, etc.
9. Ruggedness and accessibility of different regions to be accounted for places accessibility.
10. Landscape uniqueness and remarkable observations which attract visitors and tourists.
11. Participation in the national development, and this reflects the development and success of NEOM project and its benefit for the Kingdom of Saudi Arabia as a whole.

Therefore, the selected implements for SLM compatibility for NEOM Region can be illustrated in a matrix form as in Table 9.1.

## 9.3 Urban Sites

When new urban areas, clusters, sites or settlements are planned to be established; however, there is always a contradictory in describing these works. Thus, it can be assigned as urban planning or commonly used as "urbanism" where the latter represents a simulation system for the analysis of urban development that incorporating the integration with several systems including land use, transportation, the socioeconomic, and the environment aspects.

Urban sites can refer to towns, cities or suburbs whether they are located in urbanized or rural regions. Also, an urban site or area includes the major town or city itself, and the surrounding areas. Hence, urban areas or settlements can be categorized into four types (according to their geographic patterns and texture). These are, isolated, dispersed, nucleated, and linear.

Usually, for new urban areas, new model systems are developed to respond to the emerging requirements (economic, social, environmental, etc.), and has now been applied in several metropolitan areas (Waddell 2002). Therefore, to plan for new urban areas, there must be a number of considerations to be taken. These largely

**Table 9.1** Matrix for SLM compatibility in NEOM Region

| SLM Implement | Compatibility in NEOM Region | | | | |
|---|---|---|---|---|---|
| | VC | C | MC | SC | NC |
| Comply with national policies | | | | | |
| Gross land use compatibility | | | | | |
| Defining the hazardous zones | | | | | |
| Zones with natural resources | | | | | |
| Heritage and archeological sites | | | | | |
| Infrastructures and energy sources | | | | | |
| Geography of the proposed projects | | | | | |
| Climatic conditions | | | | | |
| Ruggedness and accessibility | | | | | |
| Landscape uniqueness | | | | | |
| National development Participation | | | | | |

*VC* Very compatible, *C* compatible, *MC* moderately compatible, *SC* slightly compatible, *NC* non-compatible

include the environmental and socioeconomic ones that incorporated with proper urban planning approaches that involve future changes.

For NEOM Region, proposing new urban sites is a necessary step for the development of the entire region as a global agglomerate that will include different activates and mainly the economic clusters.

### *9.3.1 Significance of Urban Sites for NEOM Region*

It is normal and necessity to plan for establishing urban sites, with different aspects and dimensions, in a planned zone like NEOM with 26,500 km$^2$. The planning and proposing of these sites need a detailed studies and investigations taking into accounts the temporal and spatial dimensions. This requires conducting field surveys to analyze factors influencing land use and the natural cover components as well as deducing the feasibility and identify needs and the expected changes. This is utmost significant but it is can be considered as a feasible and easy-doing projects, because the region of NEOM is still bare and, except the natural constraints, no other constraints can make a barrier to develop urban projects in this region.

The majority of necessity (and motivations) to establish urban sites in NEOM Region can be summarized as follows:

1. The decentralization of development of NEOM Region, and thus urban sites and the related human activities must uniformly cover the entire region where it is necessary to exist.
2. From the national point of view, Saudi Arabia needs to have a balanced urban expansion in all regions. Thus, NEOM Region is still bare area with very fine urbanism. Therefore, the prosed urban sites will open the chance for this expansion.
3. The proposed projects and activities in NEOM Region need to mobilize large number of people who will permanently stay there. These people need a homes and residences where they can settle in.
4. The expatriates with short-term stay (e.g. visitors, tourists, etc.) will demand comfortable dwelling for their stay. This can be reached if residence sites (e.g. hotels, restores, etc.) are constructed.
5. There are commercial centers and companies need to have localities for their works. This also can be included as a part of the proposed urban areas.

According to Litman (2019), the main geographic areas are often classified as follows:

- Village—relatively small urban site (almost less than 10,000 people).
- Town—relatively medium size urban site (usually less than 50,000 people).
- City—large settlement (almost exceeds 50,000 people).
- Metropolitan—it is a large urban region (usually exceeding 500,000 people) and almost including one or two large cities.
- Central business district (CBD)—represents the main commercial center in a town or city.

### *9.3.2 Proposed Urban Sites*

As mentioned in the significance of establishing urban sites, the proposition of these sites need in-depth investigation and development analysis which can be applied for selected zones form NEOM Region. Nevertheless, the following sites can be primarily proposed as in Table 9.2, but more sites can be proposed once an integrated urban plan is adopted for the region as a whole.

It is obvious that there is a miscellany of urban sites and the related activities needed for NEOM Region. These can be implemented over different time periods (including prepared timeframe) which is controlled mainly by the project needs, priorities and engineering applications and even the future vision. In addition, the geographic distribution of these sites must be initially determined. However, the majority of these sites implies the coastal zone, as well as the coast itself and the

**Table 9.2** Proposed urban sites in NEOM Region

| Type | Examples | Location[b] | Estimated number/dimension[c] |
|---|---|---|---|
| Resorts | Coastal resorts and residences | c | 1/10 km along the coast |
| Hotels | Hotels with different levels | c, cp | 3/25 $km^2$ |
| Urban building | Compound homes | cp, pd | Cluster/10 $km^2$ |
| Mountain homes | Small-scale homes | mr | 1/100 $km^2$ |
| Medical centers | Hospitals, laboratories, etc. | all | Selected |
| Commercial business district (CBD) | Building for trading, finance, investment and other companies, etc. | cp | 10/2 $km^2$ |
| Educational hubs | Research centers (e.g. schools, universities) and fixed monitoring sites | cp, pd, pl | 1/0.25 $km^2$ |
| Smart villages[a] | Home with green energy | pd, mr | 5 in the entire region |
| Markets | Commercial sites and centers | cp, pd | 20/all cp and pd |
| Harbors | Small-scale ports for yacht and touristic activities | c | 1/10 km along the coast |
| Industrial zones | To be determined with decision makers and areas requirements | All | Selected |
| Administrative offices | Official and governmental sites (municipalities, police offices, etc.) | All | |
| Distinguished urbanized landmark | The Island of the Kingdom (Fig. 9.2) which can be well observed at a distance from space | c | One landmark |

[a]Smart villages will be attributed to green energy section
[b]*c* Along the coast, *cp* coastal plain, *pd* piedmont, mountainous region, *all* all regions, *pl* plateau
[c]As per the first 5 years of the project

piedmont region at the contact between the mountainous region and the coastal plains (Table 9.2). Thus, the coastal region is supposed to occupy the main urban sites.

The dimensional aspects are very crucial and should be also determined in the earlier stages. In other words, the number of sites to be established should be determined within a specific time period. Here, in this preliminary proposition the proposed sites are sought for the next 5 years (up to 2025) as a first work stage. This can be tentative and flexible according to the project management plans and the financial resources available as well as the availability of foreign and local partnership in the project. Nevertheless, the author is just creating a scheme for compatibility for land management as a whole.

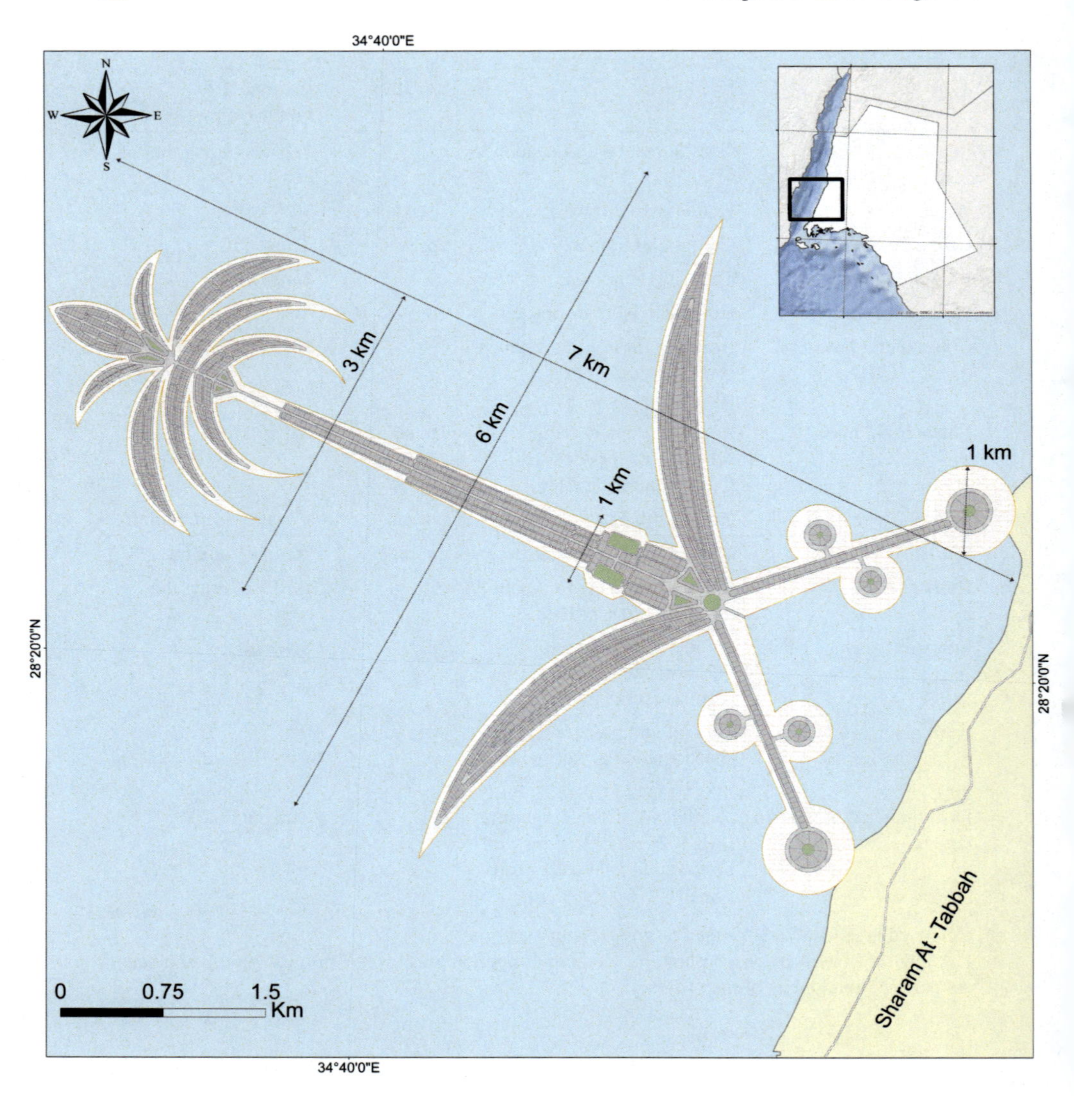


**Fig. 9.2** Island of the Kingdom, a proposed coastal urbanized landmark for NEOM Region

## 9.4 Transportation

Transportation in urban development represents the veins connecting different urban sites, and it comprises an urgent aspect of urban planning. Thus, transportation in land management composes a key issue for sustainable development. The highly mobile transportation system has affected land use patterns, particularly how people choose to locate their homes and businesses. Conversely, spread out land use patterns further increase the demand for transportation because of remote travel distances. That is why transportation is one of the six principal components of land use.

Thus, major transportation routes are usually found before the construction of urban sites and it is developed later on to connect between the minor urban sites. While, transportation is sometimes built on after the presence of urbanism.

### 9.4.1 Significance of Transportation NEOM Region

Transportation networking is urgent for NEOM Region. This is because the existing transport networks will be insufficient for the region when it becomes an international commercial site. Therefore, NEOM Region has only three main highways span from Tabuk Region (NE-SW direction) towards the coastal zone where these three transport routes join connect Saudi Arabia with Jordan at the Gulf of Aaqba. While, there no railway or flight ports can be considered up to date in NEOM Region, except two runways whereas one of them is proposed to be the airport of the region. Thus, the following is the significance to develop a complete transportation system for NEOM Region:

1. NEOM Region, as the northern gate of Saudi Arabia, is still without complete transportation system.
2. Reaching NEOM Region by flight is still tedious because the nearest airport is located in Tabuk City which is approximately 200 km from the coast.
3. Even though Duba Port is the major site for trading and commercial purposes close to NEOM Region, yet the region does not have a railway network to support these purposes.
4. The existing three highways, which extend diagonally from SE (from Tabuk and the surrounding) to the coast, do not have lateral connections between each other and this retards the joining between these three highways and then leave the regions between them unreachable.
5. When NEOM Region becomes ready for different operational activities, domestic transport networks will be necessary to facilitate the movement in the urbanized areas (i.e. smart growth).

### 9.4.2 Proposed Transportation for NEOM Region

It is urgent to have a complete transportation system in NEOM Region that connecting the region with the national and international places, as well as to networking all properties within the region.

For roads network, there should be a detailed transportation scheme made specifically for the entire NEOM Region, and this can be done within a context of an integrated planning and land management approaches. This must be built along with the specified localities for the proposed urban sites. For this reason, roads network will not be illustrated, and therefore, only rail transport and air transport will be proposed in this document.

1. Rail transport:

This represents a safe, most dependable and comfortable land transport system if compared to other forms of transportation. It covers large distance at high speed and with cheaper mode of transport.

Based on the transportation requirements for NEOM Region; however, three aspects of rail transport can be proposed as shown in Fig. 9.3.

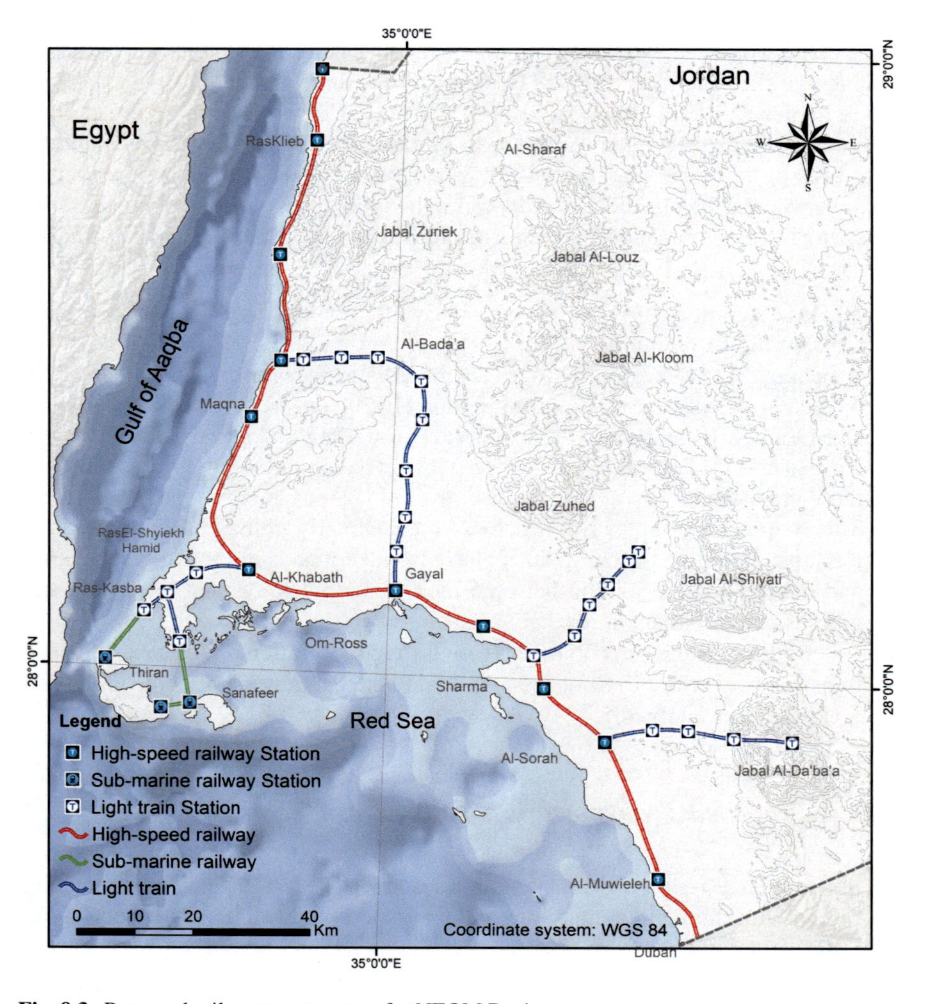


**Fig. 9.3** Proposed rail transport system for NEOM Region

- High-speed railway: These are trains with high speed and can serve travels for long distances (i.e. hundreds of kilometers). They are proposed to facilitate transportation from/to NEOM Region as well as they can be as trans-border trading and travel tool.
  The proposed route of this railway is shown in Fig. 9.3. It is about 200 km long (within NEOM Region), where it has one coastal route with 11 stop-stations along its entire way. Thus, it can continue after NEOM Region for more than 45 km along Haqel, and then to the Jordanian border at Al-Durrah border crossing.
- Light train: This type of railway are characterized by moderate to slow speed (50–100 km/h). Thus, light train is used for short to moderate distances and usually its stations are connected with the high-speed trains.
  Light-speed trains are used to reach minor localities, and this is why they were proposed for NEOM Region where 5 routes of light-speed trains are derived from the costal line inside the mountainous areas (Fig. 9.3).
  The total proposed distance of the light-speed train for NEOM Region is about 165 km where 22 stop-stations were proposed.
- Sub-marine train: Usually sub-marine trains are used when a train route is cut by water (e.g. river, lake, sea, etc.). However, sub-marine train can be also used as a touristic tool to reach marine places and to enjoy observing underwater views and this is the compound objective of proposing sub-marine trains for NEOM Region.
  The proposed train connects between land and near Ras -Kasba, Ras El-Shyiekh Hamid with Thiran and Sanfeer Islands where a round route of about 30 km is proposed including 3 stop-stations (Fig. 9.3).

2. Air transport

The proposed air transport in this document implies proposing landing sites and not flight routes. Hence, for NEOM Region there must be a major airport and helicopter landing sites. This can be as follows:

- International airport: As mentioned previously there are two runways; one in Wadi El-Masier (2.1 km) and the other one near Sharma (3.25 km) and both are located along the coast. However, the length of Wadi El-Masier Runway does not fit commercial plans, while the length of Sharma Runway is also insufficient and should be 4 km at least.
  There is an obvious constraint for Sharam Runway where the location is prone for flood hazard since the entire zone there is situated along Wadi Sharma pediment. For this purpose, a schematic figure has been illustrated for a proposed international airport for NEOM Region as shown in Fig. 9.4. Hence, the proposed airport for NEOM Region possess a number of advantages where the natural hazard safety side has been primarily accounted, plus its characteristic as another distinguished landmark that can be well observed from space and can assign the region with such ma-made feature.
  The proposed airport occupies four runways with international dimensions, as well as a giant airport compound (e.g. halls, walking tunnels, restaurants, parking, etc.)

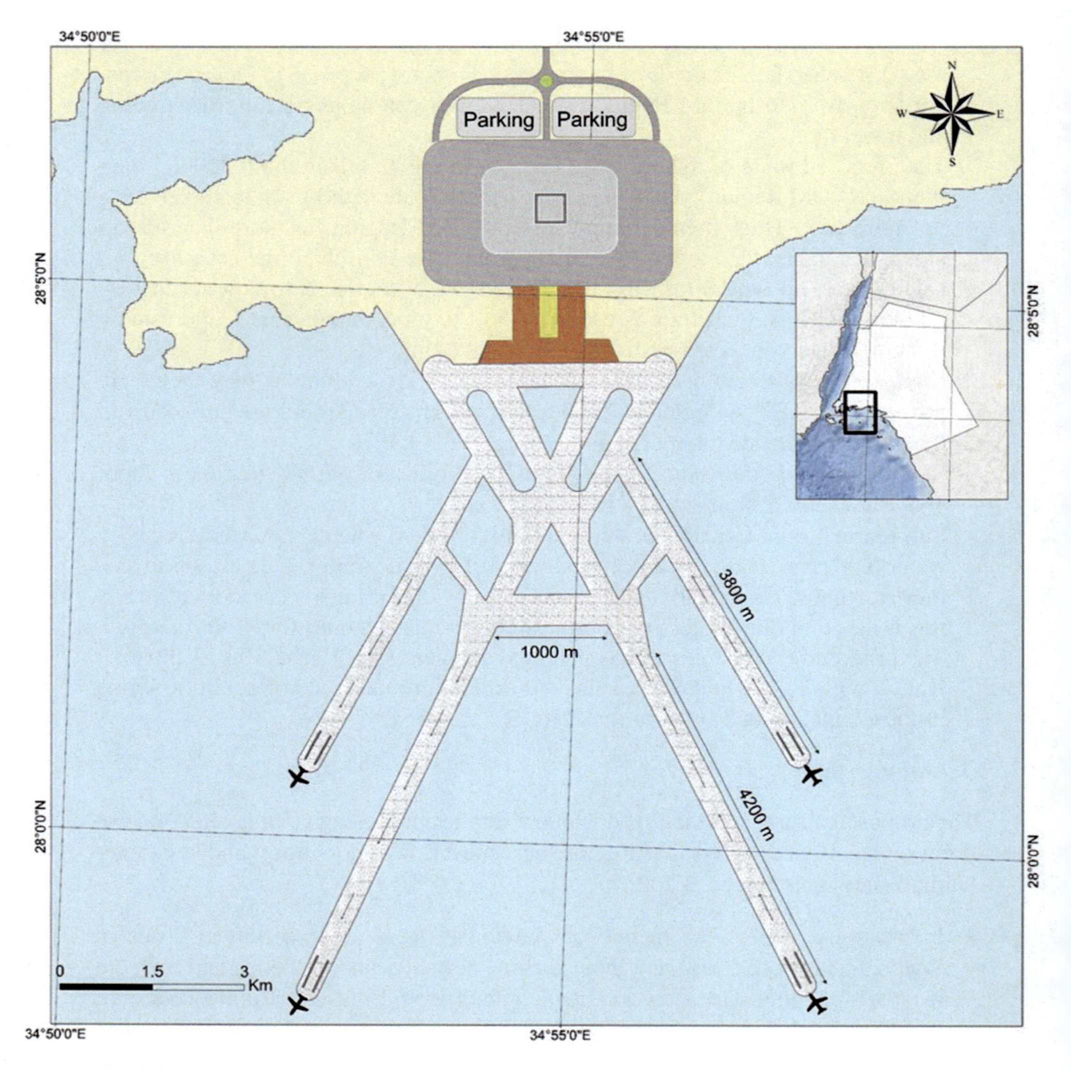


**Fig. 9.4** Proposed international airport for NEOM Region

with more than 100 km$^2$ surface area. Moreover, this airport spans its runways from the sea side, and then ends at the land area (Fig. 9.4).
For the proposed airport (as well as the Island of the Kingdom), all ecological parameters were considered as the selection of the two sites were carefully investigating in order to avoid destruction of the existing marine ecosystems.

- Helicopter landing sites: Using low-flight aircraft has become widespread and essential in many smart zones. This can be a transport tool used for many purposes. Thus, in NEOM Region, there are 8–12 helicopter landing sites can be proposed, and they can be attributed to at least 4 equipped landing stations. These stations can be distributed in selected areas considering distances, altitudes, proximity to urban sites and landmarks, etc. Thus, the following introduce the main objective of helicopter landing sites:

1. Tourism purposes, in order to reach remote and rugged sites.
2. Air ambulance to move patients to and from healthcare facilities, notably during accidents.
3. Patrols and security missions.

## 9.5 Green Energy

There are many environmental and socioeconomic benefits of using green and renewable energy. According to United States Environmental Protection Agency (USEPA 2017), these can be summarized as follows:

- The generated energy does not produce no greenhouse gas emissions from fossil fuels and reduces some sorts of air pollution.
- It creates a diversity in energy supply and reduces dependence on imported fuels.
- It creates economic development and new jobs in the fields of manufacturing, installation, etc.
- There should be various initiatives in NEOM Region, especially those which encourage the protection of the environment and the spread of culture and environmental awareness. Among these initiatives are the non-polluting green energy uses.

In this regard, NEOM Region occupies several resources which are favorable to be utilized as green energy. This includes mainly the wind and solar energy sources. They can provide energy with zero-carbon and lowest percentage of fossil fuel consumption, and thus clean environment can be reached.

Other than the benefits illustrated by USEPA (2017), the main benefit of using green energy in NEOM Region includes mainly the encouragement towards utilizing new adapting measures for futures changes in the physical and anthropogenic components, and more certainly the fossil fuel energy shortage, as well as the in-creased pollution rates.

NEOM Region occupies a potentiality for wind, solar, geothermal, hydropower and sand energy sources. However, to determine the feasibility of these sources, an in-depth assessment is required. Nevertheless, the most available green energy sources for the time being can be suggested as follows:

1. Wind energy

Wind energy offers many advantages, including that; it is cost-effective, clean fuel source, a domestic source of energy and it is sustainable. Nevertheless, there are some challenges in using wind power, such as that suitable wind sites are usually located in remote regions far from cities where the electricity is needed, also turbine blades could damage local wildlife and might cause noise and aesthetic pollution (EERE 2020).

For wind energy, location is the most significant factor where it should link between the level and availability of wind power and the urban sites requires green energy. Of course, this needs field investigations that integrated with the urban planning schemes for NEOM Region, including the coastal, mountainous and plateau areas.

As a preliminary proposition, there are 7 sites selected to install wind turbines in NEOM Region (Table 9.3), most of them are proposed to be on the mountains facing the coastal zone where urban sites and the related activates will take place. In addition, there can be one in the plateau area to deliver energy to the urban site there. Moreover one project is proposed to be in Sanafeer Island where it represents a typical environmental application in the region. These are shown schematically in Fig. 9.5. Thus, the selection of these sites considered the following:

- Presence of wind corridors and exposures with sufficient and continues wind activity.
- Feasibility to transfer the power for urbanized areas, which proposed to be established within the context of NEOM project.
- No disturbance or wildlife impact.

2. Solar energy

This type of green energy is flexible energy technology where it can be fixed at or near the localities of use; or as a central-station. In both cases, energy can be stored and produce for distribution after even the sun sets. No doubt that solar energy becomes an effective source of energy and widely spread in many smart projects.

NEOM Region, with good sun exposures can be considered as one of the most suitable region for solar power generation. This can be viewed from two aspects: (a) pilot projects where a large number of solar cells are installed to store energy for further delivery into urbanized areas, (b) local power plants where they can be fixed separately on homes and buildings but within an integrated green energy project.

**Table 9.3** Location of the proposed green energy sites in NEOM Region

| Green energy type number[a] | Wind | Solar | Sand |
|---|---|---|---|
| 1 | Jabal Al-Jadrieh | Al-Sharaf | Markaz Al-Sayani |
| 2 | Jabal Al-Na'ayjat | Al-Bada'a | Meshash Dumm |
| 3 | Jabal Koluib | Maqna | Abar Qana |
| 4 | Jabal Zuhed | Abar Al-Osieli | – |
| 5 | Jabal Al-Somakh | Bir Treem | – |
| 6 | Jabal Hareb | – | – |
| 7 | Sanafeer Island | – | – |

[a]Number according to Fig. 9.5

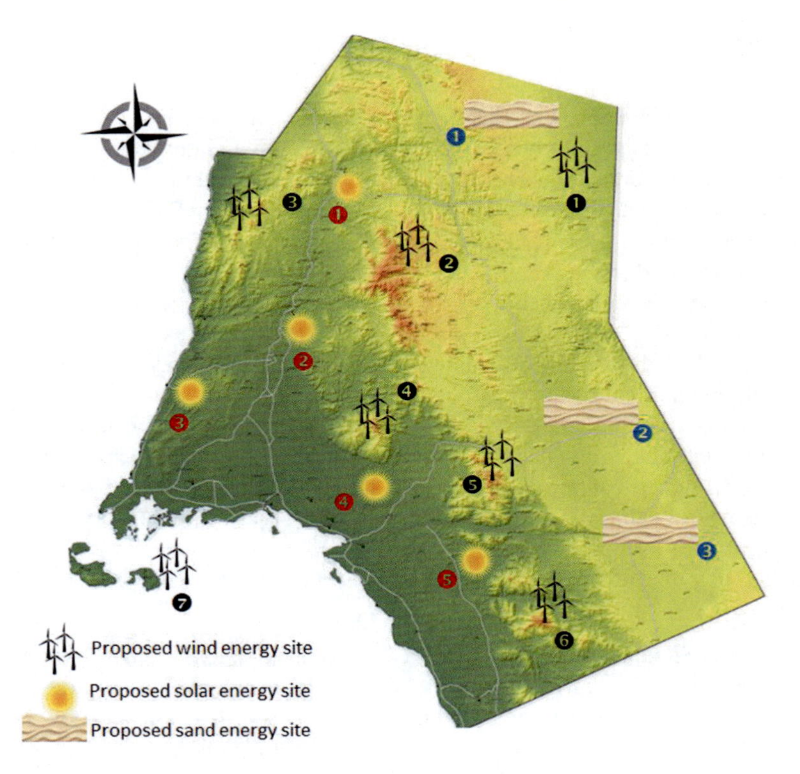


**Fig. 9.5** Schematic proposition of green energy sites in NEOM Region

There are 5 preliminary sites proposed to execute pilot projects for solar energy in NEOM Region (Table 9.3 and Fig. 9.5). There are 4 sites distributed mainly in the coastal zone and one site near Al-Sharaf town.

3. Sand energy:

The technology of using of sun-heated sand to generate energy is still in its initial stages. It is an attempting to bring together two things the desert has in abundance—sand and sunshine in order to produce power cleanly and efficiently. The idea behind is a concentrated solar power (CSP) system by using mirrors to concentrate sunlight to heat sand, which is in turn generate steam that can run turbines and generate electricity.

This technology is being attempt in UAE where desert sand can capture heat which can be stored in cells and then delivered a green source of energy (Diago et al. 2018). Thus, for NEOM Region there are 3 sites proposed (Table 9.3 and Fig. 9.5).

## 9.6 Heritages and Archeology

In June 2020, the Kingdom of Saudi Arabia won for the first time the membership of the World Heritage Committee, which carries with it an international affirmation of the distinguished position that the Kingdom has achieved in the field of concern for its national heritage. In addition, Saudi Arabia joined IFAP (information for all program) as a part of UNESCO. This shows the priority that the Kingdom has given to its historical and traditional aspects.

Similar to many regions in Saudi Arabia, Tabuk Province occupies a large number of heritage and archeological sites, where NEOM Region possess significant ones of these sites. Thus, the identification and development of these sites will attract people from different parts of the world.

The aim of focusing on these sites for NEOM Region is to add a cultural value to this region and this will make it as a region with several attracting spots, including heritage and archeological ones. These site are under the mandate of Saudi Ministry of Tourism, and namely the Saudi Commission for Tourism and National Heritage (SCTH).

The following are the recognized heritage and archeological sites in NEOM Region:

- Maghayer Shouaib (or Madeen): This is located in Al-Bada'a, and it has an ancient oasis named by Ptolemy as "Al-Aynieh" where there are several graves sculptured on rocks which are attributed to the Nabataean Era. It has also a town called "Al-Mutlaqa" which was originated from an early Islamic period. The scattered ruins indicate that the region has witnessed different nations with commercial and agricultural activities which was BC.
- Al-Malha: It is an archeological site relates to the Nabatean period. It is located north of Al-Bada'a, near the Maghayer Shouaib.
- Aynounah: It is an oasis (90 km north to Duba), located on coast where it was the famous Port of Anbat (Loki Komei) or the so-called the White City. Its ruins a still exist near Al-Ain. It also contains Islamic ruins.
- Treem: An archeological site located near Al-Muwieleh.
- Kala'at Al-Muwieleh: It is a castle that was one of the main stations of the coastal pilgrim in the late Islamic ages where it was built by Sultan Suleiman. Some historians consider it as one of the largest castles in Saudi Arabia.

## 9.7 Entertainment

The entertainment, as an activity or event, is usually performed to hold attention and interest of people and it gathers people together and amuses them in their leisure time. Therefore, it believed to be a necessary task, and people often dedicate time for the entertainment in their life programs.

All smart cities and urban zones with remarkable features are usually establish entertainment localities as a supplementary implementation to support the commercial and socioeconomic activities. Thus, entertainment tourism, as a recently adopted work, encourages tourists to travel around the world and are an important part of many gaming destinations (Luo and Lam 2017).

The selection of entertainment activities in a smart zones is con-trolled by several factors, including mainly: the natural setting and cultural attributes of the area of interest, types of commercial and economical purposes to be carried out, as well as the social and cultural and even religious belonging of inhabitants. These factors were considered in this document in order propose the most appropriate entertainment activities for NEOM Region.

The majority of the proposed entertainment activities for NEO Region implies many entertainment aspects which can be implemented or put into planning programs Table 9.4.

Normally, the proposed entertainment activities can be integrated with other proposed tasks in NEOM Region. For example, entertainment can be considered as an attractive feature added to the heritage and archeological sites. Whereas, the tourism might include also visits to gold and silver mines and so on.

**Table 9.4** Proposed entertainment activities for NEOM Region

| Activity | Description | Proposed number[a] | Location[b] |
|---|---|---|---|
| Museum | National museum for natural history (collections from KSA and abroad) | 1–2 | Urbanized coastal zone where dense commercial and economic activities are proposed |
| Theaters | Central theater | 1 | |
| Libraries | Scientific, art and literatures books (national and international) | 2–4 | |
| Educational center | Education assemblies for knowledge mapping | D | |
| Music centers | Learning and exposing musical activities | 1–2 | |
| Cinemas | Media tool highlights on different topics including community issues | 3–5 | |
| Amusement park | For various attractions for all ages (kids to elders) | 3–5 | |
| Gardens | Comfortable landscapes for gatherings | 10–15 | Different places |
| Sports city | Stadium and other sport activities (e.g. horse and car racing, etc.) | 2 | |

(continued)

**Table 9.4** (continued)

| Activity | Description | Proposed number[a] | Location[b] |
|---|---|---|---|
| Mountain sports | Handrail and sand vehicles, mountain climbing etc. | 8–10 | Selected mountainous sites |
| Snow sites | Snow cover entertainments | 1–2 | |
| Balloons | Tourist balloons | 2–3 | Different places |
| Zoo | National zoo (local and international species) | 1 | Mountainous region |
| Geological sites | Distinguished geological exposures (e.g. fossils, ores, etc.) | 10–12 | |
| Restaurants | Rest places and feeding purposes | D | Different places |
| Yachats | Yachats centers for tourism and research purposes | 8–10 | Selected sites on shoreline and the islands |

[a]Proposed number for the entire NEOM Region
[b]Selection of locations remains dependent on many factors and the existing constraints
*D* It depends on area requirements

# References


AAII (American Association of Individual Investors). (2020). *The top 10 economic indicators: What to watch and why*. Available at: www.aaii.com/investing-basics/article/the-top-10-economic-indicators.

Diago, M., Cresponlniesta, A., Soum-Glau, A., & Calvet, N. (2018). Characterization of desert sand to be used as a high-temperature thermal energy storage medium in particle solar receiver technology. *Applied Energy, 216*, 402–413.

Dumanski, J., Gameda, S., & Pieri, C. (1998). *Indicators of Land Quality and Sustainable Land Management. An annotated bibliography. Agriculture agri-food Canada* (p. 134). Washington, D.C.: The World Bank.

EERE (Office of Energy Efficiency& Renewable Energy). (2020). *Advantages and challenges of wind energy*. Available at: https://www.energy.gov/eere/wind/advantages-and-challenges-wind-energy.

Hammond, A., Adriaanse, A., Rodenburg, E., Bryant, D., & Woodward, R. (1995). *Environmental indicators: A systematic approach to measuring and reporting on environmental policy performance in the context of sustainable development* (58 p). World Resources Institute.

Litman, T. (2019). *Evaluating transportation land use impacts considering the impacts, benefits and costs of different land use development patterns* (72 p). Victoria Transport Policy Institute. Available at: https://www.vtpi.org/landuse.pdf.

Luo, J., & Lam, C. (2017). *Entertainment tourism. Social sciences, tourism, hospitality and events* (1st Ed., 76 p). London: Routledge. doi https://doi.org/10.4324/9781315162652.

Sam, K., Coulon, F., & Prepich, G. (2016). Working towards an integrated land contamination management framework for Nigeria. *Science of the Total Environment, 571*. doi 10.1016/j.scitotenv.2016.07.075.

USEPA (United States Environmental Protection Agency). (2017). *Local renewable energy benefits and resources*. Available at: https://www.epa.gov/statelocalenergy/local-renewable-energy-benefits-and-resources.

Waddell, P. (2002). Urbanism: Modeling urban development for land use, transportation and environmental planning. *Journal of the American Planning Association, 68*(3), 297–314. https://doi.org/10.1080/01944360208976274.

# Chapter 10
# Discussion and Conclusion

**Abstract** The Kingdom of Saudi Arabia, as a leading Arab and Islamic country, is well known by its economic achievements during the last few decades. This rose the level of the Kingdom among the international assemblies and made it as a pioneer in the economic, industrial and commercial levels. Hence, it is rarely a year pass by in Saudi Arabia without the announcement of a giant project to be implmented on national development where the sustainability is always considered in these projects. It might not the last one, the "NEOM Region" project does not account only for the national level, but for the global level. It has been come to light during October 2017 and works have been already started there to change this bare region with rugged mountains, vast desert sands and empty coast to the "Dream Zone" and the global spot that attracts people from the entire world to invest in this region and enjoy their time in a place with uniqueness for its landscape, climate, traditions, resources and the biological species and biodiversity. However, the project still lacks for dessimination. Thus, the idea of this book aimed at producing a document with combined themes on NEOM Region. Hence, the author presents this book for concerned stakeholders and high-level decision makers. It is a first document of its type, but further and in-depth development assessment is still needed.

**Keywords** Global hub · New vision · Commercial zone · Sustainable development · Book on NEOM Region

## 10.1 Promotion of Results

There are several outcomes and finding resulted from this documents implying a miscellany of themes and subjects. Hence, the results were introduced by different dissemination aspects including text discussion, tables, photos, maps and figures. This facilitate using these data and understand by different t stakeholders including experts and decision makers.

The achievement of NOEM Project will be carried out by several entities (e.g. companies, establishments, expert teams, etc.) where each of these entities is specialized in defined theme and specific type of project execution. Therefore, the present



M. M. Al Saud, *Sustainable Land Management for NEOM Region*,
https://doi.org/10.1007/978-3-030-57631-8_10

document can be handled for these themes to build an overall idea about the region, notably that the region is still lack to comprehensive document and maps to reveals its major features and components which are need for projects execution.

No doubt that all themes discussed in this document were to illustrate the majority of scopes that must be discussed and analysed in-depth during the execution of projects. Even though, the author mentioned several times that all data and information as well as measurements occupied in this document can be much investigated and this depends on project requirements. For example, the map of flood-prone areas covers the entire region on NEOM, but for specific localities, such as those located in the coastal zone, where many urban sites will be established, more detailed maps can be produced using the same models and tools of analysis used in this document, but with more detailed scale.

This is also the case for the other themes in the documents; and therefore, the illustrated dimensions and measurements can be also determined with much accuracy for local applications (i.e. pilot projects). This can be also achieved using similar methods of data retrieve and analysis, but with more detailed scales, where the temporal variations should be considered as well.

As one of the advantage in data preparation in this documents is the production of digital data. This includes:

1. Measurements and dimensions followed most creditable formulae and statistics. Then, they are available in tabulation form.
2. Maps and all geo-spatial data are in the digital form, and more specifically in the GIS systematic files (i.e. shape-files), and this facilitate applying different manipulation processes, as well as adding/modifying any changes that might needed.
3. Thematic data were prepared specially for NEOM Region with complied documents and not as separate hard copy sheets as most of them were found.
4. Satellite images are also available, either as processed ones or as raw data, and this assist in investigating these images for further data be extraction.
5. The produced numeric data can be manipulated and recalculated in the view of the changing conditions (e.g. climate change, increased groundwater abstraction, extinction of species, etc.).

The outcomes can be tentatively allocated according to different projects, thus book themes and their applicability for NEOM project are show in Table 10.1. Hence, some of these themes can be used for more than one application, such as (for example) the slope map can be utilized for risk assessment and for water harvesting.

## 10.2 Chapters' Highlights

The outcome of the illustrated chapters can be summarized by high-lighting on the key points of each chapter. This can be as a simplified tool to tag the major results for further review. Therefore, the following presented the highlights for each chapter:

**Table 10.1** Book themes and their applicability for NEOM project

| Theme | Major applicability | Minor applicability | Temporal variation |
|---|---|---|---|
| General discussion and data analysis | General assessment, discussion and feasibility for different theme needed | | Periodical updating |
| Physical characteristics | Principal database inventory for several applications | | |
| Topographic map | Topographic setting | Guide map | Long-term |
| Geologic map | Geological-related studies (e.g. resources exploration) | Themes where geologic is a supplementary factor | No variation |
| Selected thematic maps | Miscellany of field applications and surveys. For example DEM is used for surface water flow regime | | Long-term |
| Drainage system analysis | Water harvesting (e.g.) and flood controls | For canals, erosion assessment and water hydrology and supply | |
| Groundwater potential map | Determining groundwater availability for new projects | Water supply for pilot agricultural projects | |
| Ore maps | Metallic and non-metallic reserve identification and exploitation | | No variation |
| Biodiversity analysis | Identifying the existed species and conserving them | Ecological controls | Long-term |
| Flood map | Avoiding risk zones while planning for new human activities | Selecting the appropriate controls for risk mitigation and reduction | Short-term |
| Instability map | | | |
| Earthquake maps | | | |
| Land management compatibility | Creating geo-spatial identities for different urbanizing projects to support NEOM project and assure its success | | Periodical updating |

- Chapter 1 Highlights

- The geography of NEOM Region was highlighted as a global hub.
- Focuses were put on the project of NEOM and its motivations.
- The significance of NEOM Region has been addressed.
- Vision for future developments and SDG vision were analysed in details.

- Chapter 2 Highlights

- Components of SLM to NEOM Region were diagnosed.
- Determining the objectives on applying SLM for NEOM project.
- Constraints of applying SLM to NEOM Region were identified as well the existing constraints were also mentioned.

- Pillars of resources management and risk management for NEOM Region were determined.
- The elements and framework of the study, to produce this documents, was structured.

- Chapter 3 Highlights

- The principal physical characteristics were determined with detailed measurements and calculations needed.
- NEOM Region has been classified into three main geomorphologic units.
- The climate (rainfall and temperature) was analysed depending on both ground stations and remotely sensed products.
- All aspects of surface water were identified.
- The geology of NEOM Region has been complied and the lithological units were organized with chronological attributes.
- Rock structures were discussed in details.

- Chapter 4 Highlights

- The principles of maps production and the use of satellite images and geo-information systems were demonstrated.
- A detailed topographic map, in digital form, has been produced for the first time for NEOM Region.
- A detailed and complied geological map, in digital form, was also produced in consistency with the organized lithological units.
- A compiled landforms map was also produced in digital form.
- Surface slope map, based on the generated SRTM DEM, was established in digital form.

- Chapter 5 Highlights

- Precise cartography has be achieved for the drainage systems in NOEM Region, where all catchments of different dimensions were digitally drawn. Hence, there are 15, 7, 23 watershed were classified as major, moderate and small-scale; respectively.
- A detailed geometric analysis, using different formulae for the major watersheds, has been elaborated to diagnose the dimensional aspects of major watersheds and their relationship to water flow regime.
- Also, a detailed morphometric analysis was also applied to the major watershed up on which surface water flow and its relationship to infiltration and other hydrological parameters.

- Chapter 6 Highlights

- A general overview on the existing natural resources in NEOM Region was discussed and their management approached were also illustrated.

- A groundwater potential maps has been produced for NEOM Region where it shows three main categories: (1) High groundwater potentiality (800 m average depth; and 1625 km$^2$ area), (2) moderate groundwater potentiality (75 m average depth; and 902 km$^2$ area) and (3) low groundwater potentiality (350 m average depth; 362 km$^2$ area).
- Rock lithologies were classified according to the presence of metallic and non-metallic ores. Thus, NEOM Region occupies at least 25 major economic ore deposits.
- Digital maps for rocks containing metallic and non-metallic ore were also produced.

- Chapter 7 Highlights

- An inventory for the biodiversity in NEOM Region has been elaborated.
- Influencers and factors on biodiversity have been discussed and they were applied to NEOM Region.
- The geographic distribution and types of terrestrial plant species (106 species related to 51 families) and animal species (213 species related to 6 animal groups).
- The existing marine species were also identified whether along the Red Sea or the Gulf of Aaqba coasts. This includes: corals, fishes, Mangroves and seagrass.
- Aspects of ecosystem loss have been illustrated for NEOM Region.
- Three terrestrial and two marine natural reserves have been proposed where they occupy unique landscape and biodiversity. These reserves have a total area of about 2950 km$^2$ (11% of NEOM Region).

- Chapter 8 Highlights

- Types and methods and tools for identifying natural hazards were discussed in details.
- Therefore, models were elaborated for the data extracted from ground records, remotely sensing and field measurements along with the use of geo-information systems.
- Digital earthquake map has been produced for NEOM Region showing areas under earthquakes risk. It was resulted that the very high and high risk is 5616 which is equivalent to about 21%.
- Digital flood-prone area maps was also produced for NEOM Region. The produced map shows that about 6% of NEOM Region is prone for flood. Whereas flood risk is concentrated in the coastal zone, and specifically where piedmonts and wadis outlets exist.
- Digital instability map was again done for NEOM Region. Therefore, the produced map reveals that 14,562 km$^2$ (55%) of NEOM Region is characterized by high terrain.

- Chapter 9 Highlights

- The compatibility of land management for NEOM Region was discussed. This includes SLM indicators for NEOM Region including different aspects of indicators.
- Therefore, 5 land management activities were proposed including urban sites, transportation, green energy, heritages and archeology and entertainment activities. Amongst these activities, the Island of the Kingdom and international airport were suggested with detailed dimensional description.
- For each of these activities, there was detailed illustration on their compatibility, significance and geographic suitability.

## 10.3 Recommendations

The following recommendations represent the author's outlook which is based on the obtained work in order to produce this book. Whereas, the author stated that there can be still detailed investigations and analysis on many of the present themes. Therefore, the author is always ready to cooperate on these investigations aiming at providing all scientific consultations and methods, as well as in converting all recommendations into executive implements.

1. A detailed land management plan should be elaborated for NEOM Region where this plan should be converted into large-scale planning (for the entire region) and small-scale planning (for identified areas) based on the requirements for NEOM project.
2. There are several consideration to be accounted for the planning approaches for NEOM Region. The most significant ones are the:
   - The sustainability of the executed works over define time periods (e.g. couple of years, decade, etc.).
   - The environmental aspects must be always included while implementing the project.
   - The ecological aspects should be preserved.
   - Natural resources must not be depleted.
3. Adoption of Public-Private- Partnership (PPP) in the workability of the project and during the sustainability periods. This will reinforce the collaboration between the governmental and private sectors.
4. Investment plans over define timeframe must be adopted, by different themes and projects, whether with local or foreign partners or both.
5. Monitoring and measuring systems and stations should established, notably that NEOM Region has very little types of monitoring systems and stations and some others ones are totally lacking. This can involve (but not limited):
   - Applying a number of urgent surveys (e.g. aerial surveys, geophysical, bathometry, ecological, etc.).

- Monitoring stations should be installed for different measurements and controlling points. For example, fixing stations to measure stream flow, meteorological variables, evapotranspiration, etc.
- Remotely sensed monitoring systems, such are Early Warning System, Space Observatory System, etc.

6. Performing a comprehensive mapping project which can be integrated with the mentioned surveys. This can include the following maps (but not limited) with fine scale:

   - Land cover/use map.
   - Soil map.
   - Geomorphology map.
   - Landform map.
   - Agriculture map.

7. Detailed biodiversity survey and classification for plants and animals, including their species and geographic distribution. In addition, there must be a number of natural reserves where the proposed ones were selected on ecological and environmental basis.
8. Selecting areas with recurrent (& obvious) natural hazard criteria (i.e. hot spots) for detailed natural risk analysis and managements with a special emphasis on earthquake, flood and terrain instability hazards. In addition, all planned sites (e.g. urban sites, roads, etc.) should consider natural risk assessment before works take place.
9. The proposed urban sites, transportation schemes, green energy sites and entertainments activities can be considered as guide lines for similar works which are essential for NEOM Region.
10. The existing heritage and archeological sites must be well reserved and put under the mandate of the Saudi Commission for Tourism and National Heritage (SCTH).

# About the Author

**Professor Mashael Bent Mohammed Al Saud**
is a Research Scientist at the Space Research Institute, King Abdel Aziz City for Science and Technology (KACST), Kingdom of Saudi Arabia. Prof. Al Saud is specialized in Applied Geomorphology and Geodesy. Hence, the Author worked in several related disciplines where space techniques and geo-information systems were utilized. Therefore, the Author involved in research studies and projects on: watershed management, applied geomorphology and surficial processes, natural hazards, water resources assessment and exploration, geophysical prospecting and many other topics where satellite images are used to monitor Earth' surface.

Prof. Al Saud is a member in numerous national and international scientific societies, and has been granted a number of national and international Awards and Honorariums. The Author has remarkable number of published scientific works including: books, publications in International Peer Reviewed Journals & Regional and International Conferences and Symposia.

Prof. Al Saud is continuously producing research projects, technical repots and thematic maps for different regions from the Kingdom of Saudi Arabia, and always aiming to add scientific inputs for the research and development in the Kingdom.



M. M. Al Saud, *Sustainable Land Management for NEOM Region*,
https://doi.org/10.1007/978-3-030-57631-8

# Index




M. M. Al Saud, *Sustainable Land Management for NEOM Region*,
https://doi.org/10.1007/978-3-030-57631-8

**N**



**O**



**P**

## S

**W**



**X**



**Y**



**Z**

Printed in the United States
By Bookmasters